GRAPHIS PACKAGING 7

GRAPHIS PACKAGING 7

. . . .

AN INTERNATIONAL COMPILATION OF PACKAGE DESIGN

EIN INTERNATIONALER ÜBERBLICK ÜBER DIE PACKUNGSGESTALTUNG

UN RÉPERTOIRE INTERNATIONAL DES FORMES D'EMBALLAGE

EDITED BY • HERAUSGEGEBEN VON • EDITÉ PAR:

B. MARTIN PEDERSEN

PUBLISHER AND CREATIVE DIRECTOR: B. MARTIN PEDERSEN

BOOK PUBLISHER: CHRISTOPHER T. REGGIO

EDITORS: CLARE HAYDEN, HEINKE JENSSEN

ASSOCIATE EDITOR: PEGGY CHAPMAN

ART DIRECTOR: VALERIE MYERS

PRODUCTION ASSISTANT: CONSTANTINE FRANGOS

PHOTOGRAPHER: ALFREDO PARRAGA

GRAPHIS INC.

(OPPOSITE) TAKU SATOH / FEC. MAKEUP SERIES

CONTENTS

INHALT

SOMMAIRE

REMARKS

ANMERKUNGEN

ANNOTATIONS

WE EXTEND OUR HEARTFELT THANKS TO CONTRIBUTORS THROUGHOUT THE WORLD WHO HAVE MADE IT POSSIBLE TO PUBLISH A WIDE AND INTERNATIONAL SPECTRUM OF THE BEST WORK IN THIS FIELD.

ENTRY INSTRUCTIONS FOR ALL GRAPHIS BOOKS MAY BE REQUESTED FROM:
GRAPHIS INC.
141 LEXINGTON AVENUE
NEW YORK, NY 10016-8193

UNSER DANK GILT DEN EINSENDERN AUS ALLER WELT, DIE ES UNS DURCH IHRE BEITRÄGE ERMÖGLICHT HABEN, EIN BREITES, INTERNATIONALES SPEKTRUM DER BESTEN ARBEITEN ZU VERÖFFENTLICHEN.

TEILNAHMEBEDINGUNGEN FÜR DIE GRAPHISBÜCHER SIND ERHÄLTLICH BEIM:
GRAPHIS INC.
141 LEXINGTON AVENUE
NEW YORK, NY 10016-8193

TOUTE NOTRE RECONNAISSANCE VA AUX DESIGNERS DU MONDE ENTIER DONT LES ENVOIS NOUS ONT PERMIS DE CONSTITUER UN VASTE PANORAMA INTERNATIONAL DES MEILLEURES CRÉATIONS.

LES MODALITÉS D'INSCRIPTION PEUVENT ÊTRE OBTENUES AUPRÈS DE:
GRAPHIS INC.
141 LEXINGTON AVENUE
NEW YORK, NY 10016-8193

(OPPOSITE) ART DIRECTOR/DESIGNER: JEFF WEITHMAN • ILLUSTRATOR: CLINT GORTHY (FOLLOWING PAGE) PHOTOGRAPHER: CRAIG CUTLER

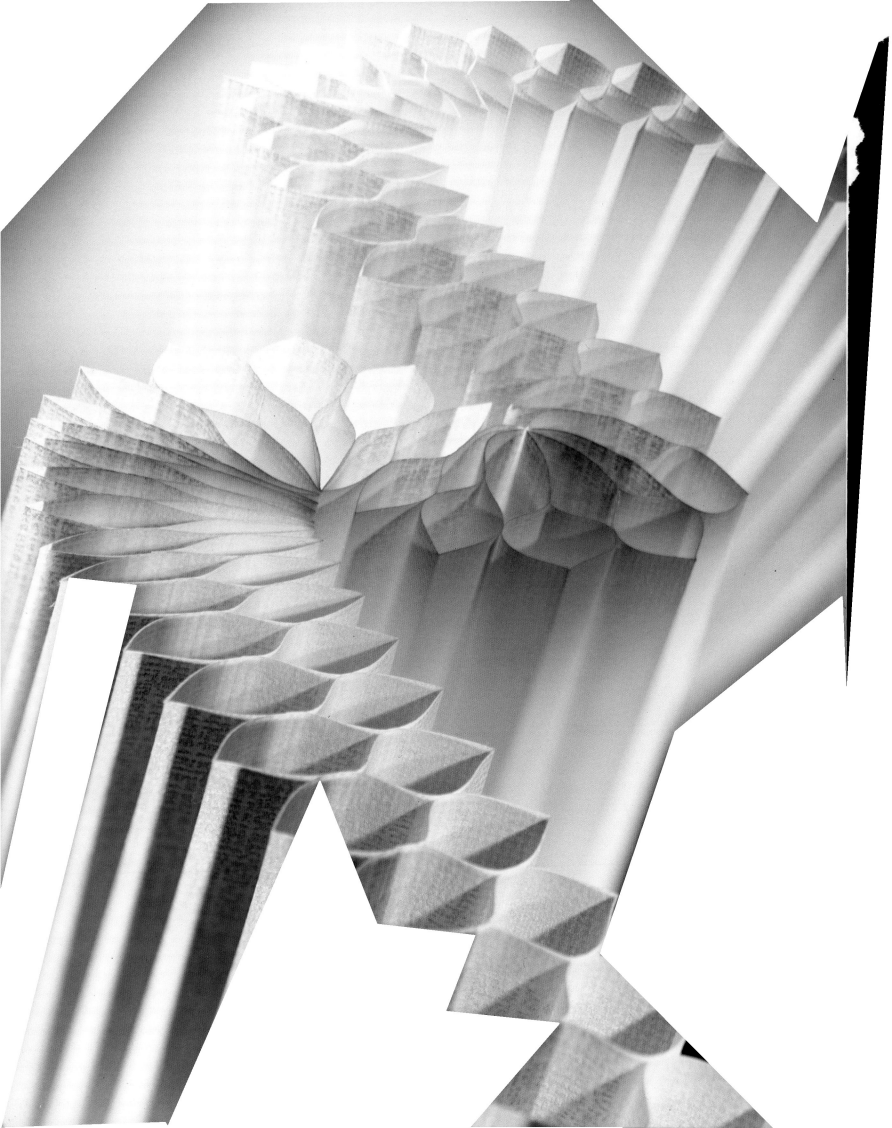

overdressed, over-packaged, glitzy presentation—far from it. In fact, in most cases the simpler the presentation, the better. A stunning package design is one created with a healthy respect for form and function. The form may be the very product itself and should be well designed anyway. The form may also be the container in which the product is held—a bottle, tin, can, carton, or box. This form, together with any label or other device which brands the product or provides relevant and mandatory information, completes the package design. ■ The complete package design must be relevant and functional, but it must also be unique and have impact. It must create a reaction of "Wow!" at the point of purchase. Intuition—a sudden bolt of creativity or a stroke of genius—takes a normal design solution to the level of absolute uniqueness and gives it this necessary impact. The continuity of a designer's personal experiences of observing people, forms, textures, and colors leads to this intuition. Most of all it comes from thinking visually, dreaming visually, and putting every idea down on paper continually and with passion. The form of a simple shape, a die-cut hole, a striking color, a surface finish, or a combination of materials may make all the difference. ■ In the creative development of packaging design, the designer must translate each product's characteristics and unique personality and style into graphic form in a succinct or eye-catching way. A successful package must naturally protect the product within, but it must also stand out in the crowded marketplace, attract attention, and sell itself. ■ Many products enter the market without the benefit of large advertising budgets or with no advertising support at all. When this is the case, the package must be absolutely sensational to succeed! It must be its own silent salesperson, attracting and convincing the consumer to purchase—often in a very short time. ■ Design has played an integral role in recent years in reshaping the marketplace within some product categories. Some very exciting and creative advances have taken place. Within the perfume market, for example, there has been an increasing number of products released with stunning packaging presentations. The wine and spirits market has also experienced major advances in quality design. This is especially the case in the "new world" wine countries of Australia, New Zealand, and the USA. As with the cosmetic industry before it, designers are developing new bottle shapes, glass colors and finishes, neck shapes, and new closures. ■ While whisky and vodka products have traditionally been at the forefront of quality packaging design in the spirits market, recently products such as Kirsch from Switzerland and Grappa from Italy have taken the lead with superbly sculptured, handblown bottles with minimal labeling. These bottles are both environmentally and aesthetically well designed. ■ The designs featured in *Graphis Packaging 7* attest to the exciting advances taking place not only in package design for the spirits industry, but for the many categories of the package design discipline.

Es ist in letzter Zeit viel über umweltfreundliche bzw. nichtumweltfreundliche Verpackungen geschrieben worden. Die Verwendung wiederverwerteter oder wiederverwertbarer Materialien und die Verantwortung des Gestalters gegenüber unserem Planeten sind ausgiebig diskutiert worden – und das zu Recht. ■ Auch ich bin der Meinung, dass der heutige Verpackungsgestalter gegenüber der Umwelt eine Verantwortung hat. Seine Pflicht ist es aber auch, eine erfolgreiche Verpackung zu entwerfen, d.h. eine Verpackung, die verkauft. ■ Jede Firma, die ein Produkt entwickelt und Stunden, Tage, Wochen, sogar Jahre in Forschung und Entwicklung investiert hat – ganz abgesehen von riesigen Geldsummen, erwartet, dass sich diese Investition bezahlt macht, sobald das Produkt auf dem Markt ist. Wenn ein Hersteller sein Produkt einem Verpackungsgestalter anvertraut, liegt die volle Verantwortung für den Verkaufserfolg des Produktes und damit des Unternehmens beim Gestalter. ■ Der sichere Weg zum Erfolg liegt in der Schaffung einer phantastischen Verpackung. Eine «phantastische» Verpackung heisst nicht eine überladene, übertriebene, glitzernde Präsentation – ganz im Gegenteil, in den meisten Fällen gilt, je einfacher die Präsentation, desto besser. ■ Eine phantastische Verpackung ist eine Verpackung, die mit gesundem Respekt vor Form und Funktion entworfen wurde. Die Form mag das Produkt selbst sein, und das sollte in jedem Fall gut gestaltet sein. Die Form kann auch das Gefäss sein, in dem das Produkt enthalten ist – eine Flasche, Büchse, Dose oder Schachtel. Diese Form, zusammen mit dem Etikett oder anderen Bestandteilen der Verpackung, z.B. das Markenzeichen oder relevante bzw. obligatorische Informationen, ergänzen die Verpackungsgestaltung. Das gesamte Design der Verpackung muss relevant und funktionell sein, aber auch einzigartig und eindrucksvoll. Die Reaktion des potentiellen Käufers beim Anblick der Verpackung sollte «Wow!» oder sonst ein Ausdruck seiner Bewunderung sein. ■ Intuition – ein plötzlicher Geistesblitz oder ein genialer Einfall – lässt eine absolut einzigartige Designlösung mit der nötigen Wirkung entstehen. Intuition entsteht durch die Kontinuität der persönlichen Erfahrungen eines Designers, die er beim Beobachten der Menschen, der Formen, Materialien und Farben macht. Am wichtigsten jedoch ist die Fähigkeit, visuell zu denken, visuell zu träumen und jede Idee zu Papier zu bringen, und zwar kontinuierlich und leidenschaftlich. Eine einfache Form, ein ausgestanztes Loch, eine faszinierende Farbe, die Beschaffenheit einer Oberfläche oder eine Kombination von Materialien mag den ganzen Unterschied ausmachen. ■ Bei der gestalterischen Entwicklung von Verpackungen muss der Designer die Eigenschaften, die Einzigartigkeit und den Stil eines Produktes auf optisch attraktive Weise in eine graphische Form übersetzen. Eine erfolgreiche Verpackung muss natürlich das darin enthaltene Produkt schützen, aber sie muss sich auch in dem Gedränge von Konkurrenzprodukten durchsetzen, die Aufmerksamkeit auf sich ziehen und das Produkt verkaufen. ■ Viele Produkte werden ohne die Unterstützung von grossen Werbebudgets auf den Markt gebracht, d.h. unter Umständen sogar ganz ohne Schützenhilfe

BARRIE TUCKER LEITET TUCKER DESIGN IN SÜDAUSTRALIEN. ER IST MITGLIED DER AGI UND DES DESIGN INSTITUTE OF AUSTRALIA. MIT DER GESTALTUNG VON FLASCHEN SOWIE MIT DREIDIMENSIONALEN DESIGNS FÜR DEN ÖFFENTLICHEN RAUM HAT ER INTERNATIONAL EINEN RUF ERWORBEN. SEINE ARBEITEN WURDEN VON DER AUSTRALIAN GRAPHIC DESIGN ASSOCIATION SOWIE MIT GOLD UND SILBER BEI DEN CLIO-WETTBEWERBEN AUSGEZEICHNET.

BARRIE TUCKER

Much has been written recently about environmentally friendly–or unfriendly–packaging. The use of recycled and recyclable materials and the accountability of designers to the planet Earth have all been debated at great length–and rightly so. I agree that today's packaging designer has a responsibility to the environment. The other important obligation that he or she has, however, is to create successful packaging design. That means packaging that sells. Any company which has developed a product and has invested hours, days, weeks, even years in research and development–not to mention huge amounts of money–expects a return on the investment when the product goes to the marketplace. Once a client has entrusted its product to a package designer, it is that designer's absolute responsibility to ensure sales success for the product and client. The guaranteed way of ensuring success for any product is to create a stunning package. A "stunning" package does not mean an

BARRIE TUCKER IS PRINCIPAL OF TUCKER DESIGN IN ADELAIDE, SOUTH AUSTRALIA. A MEMBER OF *ALLIANCE GRAPHIQUE INTERNATIONALE* AND THE DESIGN INSTITUTE OF AUSTRALIA, TUCKER IS INTERNATIONALLY RECOGNIZED FOR HIS WINE AND LIQUOR PACKAGING AND THREE-DIMENSIONAL "PUBLIC" DESIGN IMAGERY. HIS WORK HAS BEEN HONORED BY THE AUSTRALIAN GRAPHIC DESIGN ASSOCIATION AND HAS RECEIVED GOLD AND SILVER CLIO AWARDS.

COMMENTARIES

KOMMENTARE

COMMENTAIRES

durch Werbung. Wenn das der Fall ist, muss die Verpackung absolut sensationell sein. Sie muss ihr eigener stiller Verkäufer sein, der die Konsumenten anzieht und sie zum Kauf überredet – und das in sehr kurzer Zeit. ■ In einigen Produktkategorien hat Design in den letzten Jahren dazu beigetragen, den Markt zu verändern. Einige sehr aufregende, kreative Fortschritte wurden gemacht. Auf dem Parfummarkt zum Beispiel gibt es immer mehr faszinierende Produkte in atemberaubenden Verpackungen. Und auch im Bereich der Flaschenausstattung für Weine und Schnäpse hat sich das Niveau sehr gesteigert. Das trifft besonders auf die neuen Weinländer wie Australien, Neuseeland und die USA zu. Wie schon

zuvor in der Kosmetikbranche entwickeln die Designer neue Flaschenformen, Glasfarben und Oberflächen, Halsformen und neue Verschlüsse. ■ Whisky und Wodka waren von jeher führend, was das Gestaltungsniveau im alkoholischen Getränkemarkt angeht. In letzter Zeit jedoch führen Produkte wie Kirschwasser aus der Schweiz oder Grappa aus Italien durch hervorragend geformte, handgeblasene Flaschen mit einem Minium an Etiketten das Feld der Konkurrenzan. Diese Flaschen sind im Hinblick auf Umwelt und Ästhetik ausgezeichnet gestaltet. ■ Die für *Graphis Packaging 7* ausgewählten Beispiele zeigen, welch aufregender Bereich des Graphik-Designs die Packungsgestaltung sein kann.

Ces derniers temps, packagings écologiques ou, au contraire, «polluants» sont souvent au cœur d'un débat qui a fait couler beaucoup d'encre. L'utilisation de matériaux recyclés ou recyclables, la responsabilité des designers envers notre planète sont des thèmes récurrents, et à raison. Pour ma part, je suis également d'avis que les designers de packagings ont aujourd'hui une responsabilité vis-à-vis de l'environnement. Mais leur devoir est aussi

de créer des packagings réussis, des packagings «vendeurs». ■ Toute entreprise qui développe un produit et investit à cette fin des heures, des journées, des semaines voire des années de travail, sans parler du coût d'une telle opération en terme de financement, attend que ses investissements rapportent sitôt le produit lancé sur le marché. Lorsqu'un designer se voit confier le packaging d'un produit, il porte toute la responsabilité du succès de vente et donc celui de l'entreprise sur ses épaules. Le succès est assuré si ce bon génie crée un packaging... génial. Notons à ce titre que packaging «génial» ne veut pas dire présentation clinquante, exagérations outrancières ou autres excentricités? Bien au contraire, dans la plupart des cas, plus la présentation est simple, plus c'est.réussi. ■ Un packaging génial est un packaging qui respecte forme et fonction. La forme peut être donnée par le produit lui-même, et son design doit être bon. La forme peut aussi être le récipient dans lequel le produit est conservé, une bouteille, une canette, une boîte de conserve... Cette forme, alliée à l'étiquette ou à d'autres éléments de packaging, tels que la marque de fabrique ou d'autres informations importantes, complète le design. Globalement, le design d'un packaging doit être fonctionnel, mais aussi séduisant et original. Son apparence doit immédiatement accrocher l'acheteur potentiel, lui arracher un «woahh!!» admiratif. ■ L'intuition – un trait de génie soudain, une idée lumineuse – peut transcender une solution banale en quelque chose d'unique, d'absolument génial. L'intuition naît des expériences personnelles que fait en permanence le designer en observant les gens, les formes, les matériaux, les couleurs. Mais le

plus important, c'est la capacité de penser, de concevoir de manière visuelle, de rêver de manière visuelle et de coucher sans relâche, avec passion, chaque idée sur le papier. Une forme simple, une découpe originale, une couleur flash, la beauté d'une surface ou la combinaison des matériaux choisis peuvent faire toute la différence. ■ En matière de création de packagings, le rôle du designer consiste à traduire les qualités, les spécificités et le style du produit en une forme graphique plaisante à l'œil. Outre le fait de protéger son contenu, un packaging réussi doit se démarquer de la multitude des produits concurrents, attirer l'attention du consommateur et lui faire acheter le produit. ■ Nombre de produits sont commercialisés sans le soutien de gros budgets publicitaires, voire sans publicité du tout. Dans ce cas, le packaging doit être absolument génial. Il doit se vendre par lui-même, attirer le consommateur et l'inciter à l'achat, et ce dans un délai très bref. Pour certaines catégories de produits, le design a contribué à changer le marché au cours des dernières années. En termes de créativité, des progrès très intéressants ont été faits. Le marché des parfums est sans doute l'exemple le plus frappant, et l'on trouve de plus en plus de produits présentés dans des flacons somptueux. Les bouteilles de vins et de spiritueux ont également bénéficié de cet élan créatif. C'est notamment le cas des vins australiens, néo-zélandais et américains. Comme ils l'ont fait pour l'industrie cosmétique, les designers innovent, créent de nouvelles formes de bouteilles, structures et couleurs de verre, revisitent les cols, les goulots et les systèmes de fermeture. ■ La palme revient sans conteste au whisky et à la vodka qui, de tout temps, ont marqué le marché des spiritueux. Depuis peu, des alcools comme le kirsch suisse ou la grappa italienne se démarquent par l'élégance de leurs bouteilles en verre soufflé à la bouche et sur lesquelles l'étiquette, presque superflue, joue la discrétion. Ces bouteilles satisfont de surcroît parfaitement aux nouvelles exigences écologiques et esthétiques. Les exemples présentés montrent à quel point le design de packagings est un domaine fascinant.

(ABOVE LEFT AND CENTER) DESIGN FOR SEPPELT, SOUTHCORP WINES. (ABOVE RIGHT) CHRISTMAS BRANDY BOTTLE DESIGN FOR AVON GRAPHICS. ART DIRECTOR/DESIGNER: BARRIE TUCKER. AGENCY: TUCKER DESIGN.

BARRIE TUCKER EST DIRECTEUR DE TUCKER DESIGN, AUSTRALIE. IL EST MEMBRE DE L'AGI ET DU DESIGN INSTITUTE OF AUSTRALIA, ET IL JOUIT D'UNE RÉPUTATION INTERNATIONALE POUR SES CRÉATIONS DE BOUTEILLES ET SES TRAVAUX DE DESIGN TRIDIMENSIONNEL RÉSERVÉS AUX ESPACES PUBLICS. SES RÉALISATIONS ONT ÉTÉ HONORÉES PAR L'AUSTRALIAN GRAPHIC DESIGN ASSOCIATION, ET IL A REMPORTÉ LES PRIX OR ET ARGENT LORS DES CONCOURS CLIO.

PAULA SCHER

I am not an expert on package design, and I've never actually designed a package in the engineering sense of the term. I've only designed graphics to cover pre-existing packages. I agree with the opinion that the best-designed package is the egg. I'm not particularly concerned about packaging engineering, but I am very interested in shopping and if everything were packaged like an egg, I wouldn't go shopping very often.   Mostly I go shopping when I want to go out, for distraction, for sport, for entertainment, and sometimes because I need something–which is when I like shopping the least. Looking at the packaging is great fun. I actually learned typography from old packages I bought in flea markets and antique stores. Old tobacco cans, cold cream jars, milk bottles, and sardine tins supplied me with years of inspiration.   I am inspired by the current marketplace, too. I love the excitement of coming across something new, interesting, surprising, tactile, amusing, or smart. The European

PAULA SCHER IS A PARTNER AT PENTAGRAM DESIGN IN NEW YORK. SHE HAS DEVELOPED A WIDE RANGE OF SUCCESSFUL IDENTITY SYSTEMS, PROMOTIONAL MATERIALS, PACKAGING AND PUBLICATION DESIGNS. HER WORK CAN BE FOUND IN THE COLLECTIONS OF THE MUSEUM OF MODERN ART IN NEW YORK, THE MUSEUM FÜR GESTALTUNG IN ZÜRICH, THE DENVER ART MUSEUM AND THE CENTRE GEORGES POMPIDOU IN PARIS. SHE IS A MEMBER OF THE ALLIANCE GRAPHIQUE INTERNATIONALE, THE AMERICAN INSTITUTE OF GRAPHIC ARTS AND THE EDITORIAL BOARD OF THE JOURNAL OF BRAND MANAGEMENT IN LONDON.

clothing store DIESEL is especially entertaining. They provide terrific labels, wonderfully designed T-shirts and a hip, funny magazine to set the overall tone of the experience. ■ The "packaged" retail environment in which product packaging and overall identity are considered together is very appealing. When this kind of care and consideration to overall identity exists, it tends to carry over to the service in the store. Intelligent and imaginative design of identity and packaging is nothing more than intense attention paid to small details. The best retail experience is made up of a plethora of details all working together to make the shopping experience easier and more entertaining for the consumer. ■ I find that most packaging I dislike is insulting and inconsiderate to the consumer. I hate a package that's all air and no product, or has so much paper attached to it that it fills a medium-sized trash can. I don't like food packaging on which one can't read the ingredients. I resent the insulting graphics that are inevitably found on cosmetic or household products in discount chain stores. They are usually coupled with tacky half-price banners positioned in star bursts. Discounters demand this form of graphic language as a designator of consumer value. Here "value" is a euphemism for the spending capabilities of lower-middle income families. The belief is that the consumer of discount products will assume that something that is well designed costs more than it should. Changing this assumption is the greatest challenge for the design community, but we only address the problem by teaching young designers to make fun of the discount "vernacular" in school projects, thereby reinforcing the erroneous notion that good design belongs to an elite class, or that an elite class can use the vernacular of poor people in a "groovy" manner. Money and good packaging don't have much to do with each other. The best evidence of this is in the cosmetic industry, with its sleek gilded, embossed packages and fancy-schmansy bottles. Occasionally there is an exception, such as the Jean Paul Gaultier cologne that had the audacity to come in a tin can. ■ Many of the most wonderfully designed products selected for this annual belong not to a mass market category, but are experimental products or small-run promotional vehicles. A few of the featured packages might actually be found in a grocery or department store and are produced by enlightened corporations recognizing that intelligent design is a valuable and satisfying part of the shopping experience. The shopping experience is, however, getting better all the time. Faced with stiff competition and a sluggish economy, retailers have been forced to create better environments and packaging plays a major role. ■ Packaging is environmental design. It is one of the few arenas of graphic design in which the product can't disappear onto a screen. There is nothing "virtual" about a package. It is really there in three dimensions. You touch it, feel it and if it is really well designed, you keep it. It adds to the fun and sport of shopping, one of the few non-athletic activities left that bring people outside.

Ich bin keine Expertin für Verpackungen, und ich habe noch nie eine Verpackung im technischen Sinne des Wortes entworfen. Ich habe nur Graphik für bereits vorhandene Verpackungen gemacht. Wie viele andere bin ich der Meinung, dass das Ei die vollkommenste Verpackung ist. Die technische Seite der Verpackung interessiert mich nicht besonders, aber ich gehe gerne einkaufen. Wenn allerdings alle Sachen wie ein Ei verpackt wären, würde ich nicht sehr oft einkaufen gehen. ■ Meistens gehe ich einkaufen, wenn ich ein bisschen Abwechslung, Bewegung oder Unterhaltung möchte, und manchmal brauche ich etwas Bestimmtes – dann allerdings gefällt mir das Einkaufen am wenigsten. Es macht grossen Spass, Verpackungen anzusehen. Ich habe in Sachen Typographie viel von alten Verpackungen gelernt, die ich auf Flohmärkten oder in Antiquitätenläden gefunden habe. Alte Tabak- und alte Cremedosen, Milchflaschen und Sardinenbüchsen haben mich jahrelang inspiriert. ■ Aber auch die heutigen Produkte inspirieren mich. Es ist aufregend für mich, etwas Neues, Interessantes, Amüsantes oder intelligent Gemachtes zu entdecken. Die Geschäfte der europäischen Bekleidungsmarke DIESEL sind besonders unterhaltend. Sie haben hervorragend gestaltete Etiketten, wunderschön bedruckte T-Shirts und ein witziges Magazin, in dem der ganze Auftritt reflektiert wird. ■ Das ist eine sehr attraktive Art von Ladenverpackung, zu der die Produktverpackung und der gesamte Auftritt gehören. Wenn soviel Sorgfalt und Überlegung in den Gesamtauftritt investiert werden, färbt das oft auch auf den Service im Laden ab. Intelligente, phantasievolle Gestaltung des optischen Auftritts und der Verpackung ist nichts anderes als grösste Sorgfalt bis ins kleinste Detail. Eine Vielzahl von Details bewirkt in ihrem Zusammenspiel, dass das Einkaufen für den Kunden leichter und unterhaltsamer wird. ■ Die meisten Verpackungen, die mir nicht gefallen, sind meiner Meinung nach eine Beleidigung und Missachtung des Verbrauchers. Zum Beispiel hasse ich Mogelpackungen mit viel Luft und wenig Produkt oder Verpackungen mit soviel Papier, dass sie ohne weiteres einen mittleren Abfalleimer füllen. Ich mag auch keine Lebensmittel-verpackungen, auf denen ich die Ingredienzen nicht lesen kann. Graphik von der Sorte, die gewöhnlich mit Sternen gekoppelt ist, die Preisabschläge versprechen, finde ich fürchterlich. Die Discounter verlangen diese Art graphischer Sprache als Zeichen eines speziellen Wertes für den Verbraucher. Hier ist «Wert» eine beschönigende Umschreibung für das begrenzte Budget der unteren Einkommensschichten. Man glaubt allgemein, dass der Käufer von Discount-Produkten von einem gut gestalteten Produkt annehmen werde, dass es mehr kostet, als es sollte. Dieser Annahme entgegenzuwirken ist die grösste Herausforderung der gesamten Design-Branche, aber wir scheinen das Problem nur anzupacken, indem wir jungen Graphikern beibringen, sich in Schulprojekten über die «Sprache» der Discounter lustig zu machen, womit wir die irrige Meinung unterstützen, dass gutes Design zur Elite gehört oder dass eine Elite die Sprache der armen Leute auf «lässige» Art benutzen kann. Geld und gute Verpackung haben nicht viel miteinander zu tun. Der beste Beweis sind die goldglänzenden Verpackungen mit Prägedruck und die aufwendigen Flaschen der Kosmetikindustrie. Hin und wieder gibt es Ausnahmen wie beim Eau de Cologne von Jean Paul Gaultier, das ganz frech in einer Alu-Dose daherkommt.

PAULA SCHER IST PARTNERIN VON PENTAGRAM DESIGN IN NEW YORK. DAS BREITE SPEKTRUM IHRER ARBEIT ALS GRAPHIKERIN UMFASST EINE VIELZAHL ERFOLGREICHER FIRMENERSCHEINUNGSBILDER, VERPACKUNGEN, DIE GESTALTUNG VON PUBLIKATIONEN UND PROMOTIONSMATERIAL. SIE IST MIT IHREN ARBEITEN IN DER SAMMLUNG DES MUSEUM OF MODERN ART IN NEW YORK VERTRETEN SOWIE IN DER PLAKATSAMMLUNG DES ZÜRCHER MUSEUMS FÜR GESTALTUNG, IM DENVER ART MUSEUM UND IM CENTRE GEORGES POMPIDOU, PARIS. SIE IST MITGLIED DER AGI, DER AIGA UND DES REDAKTIONSBEIRATS VON THE JOURNAL OF BRAND MANAGEMENT IN LONDON.

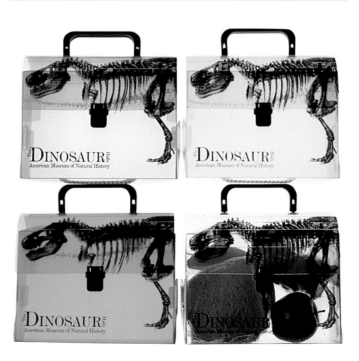

(THIS PAGE, TOP) PACKAGING DESIGN AND IDENTITY FOR UVU, A TV MODEL FROM THOMSON CONSUMER ELECTRONICS. ART DIRECTOR: PAULA SCHER. DESIGNERS: PAULA SCHER, RON LOUIE. AGENCY: PENTA-GRAM DESIGN. (CENTER) PACKAGING DESIGN FOR ÖOLA, A SWEDISH CANDY STORE FOUND IN AMERICAN MALLS. ART DIRECTOR/DESIGNER: PAULA SCHER. AGENCY: PENTAGRAM DESIGN. (BOTTOM) PACKAGING DESIGN FOR DINO STORE MERCHANDISE AT THE AMERICAN MUSEUM OF NATURAL HISTORY. ART DIRECTOR: PAULA SCHER. DESIGNERS: PAULA SCHER, LISA MAZUR, JANE MELLA. AGENCY: PENTAGRAM DESIGN.

Ich habe den Verdacht, dass die wunderschönen Verpackungen, die für dieses Buch ausgewählt wurden, nicht für den Massenmarkt bestimmt sind, sondern eher in den experimentellen Bereich gehören oder Promotionsobjekte mit kleiner Auflage sind. Es mag ein paar wenige Verpackungen geben, denen man tatsächlich in einem Lebens-mittelgeschäft oder einem Kaufhaus begegnet, hergestellt von weitsichtigen Unternehmen, die begriffen haben, dass intelligente Gestaltung ein wertvolles und befriedigendes Element des Einkaufserlebnisses ist. Das Einkaufserlebnis wird tatsächlich immer positiver, weil der Einzelhandel angesichts der harten Konkurrenz und der angespannten Wirtschaftslage ein schöneres Ambiente schaffen muss, und Verpackungen gehören zu diesem Ambiente. ■ Verpackungsgestaltung ist Umweltgestaltung. Es ist eines der wenigen Gebiete des Graphik-Designs, in dem das Produkt nicht einfach auf dem Bildschirm erscheinen kann. Bei einer Verpackung gibt es nichts «Virtuelles». Sie ist wirklich vorhanden, und zwar in drei Dimensionen. Man berührt sie, spürt sie, und wenn sie wirklich schön gestaltet ist, hebt man sie auf. Sie trägt zum Spass am Einkaufen bei, eine der wenigen nicht sportlichen Betätigungen, die die Leute noch auf die Strasse bringt.

Je ne suis pas une spécialiste du design d'emballage. En fait, je n'en ai encore jamais conçu, l'aspect technique de la chose ne m'intéressant pas particulièrement. Mon expérience dans ce domaine se limite à créer un graphisme destiné à des emballages déjà existants. Je partage l'idée communément admise selon laquelle la forme d'emballage la plus achevée est celle de l'oeuf. Pourtant, si tous mes achats se retrouvaient emballés dans des coquilles d'oeuf, il est clair que j'irais moins souvent dans les magasins. ■ En général, le shopping est pour moi prétexte à me changer les idées ou à me dégourdir les jambes. Lorsque je sors pour acheter quelque chose de bien précis, mes emplettes perdent de leur attrait, car j'ai alors moins de raisons de détailler les emballages. En matière de typographie, j'ai tout appris des vieux emballages que j'ai dénichés sur les marchés aux puces ou dans les brocantes. Ainsi, les tabatières, les pots de crème, les bouteilles de lait et autres boîtes de sardines des générations précédentes ont été pour moi une source d'inspiration constante pendant des années. ■ Cela étant, les produits actuels ne manquent pas non plus d'intérêt. Tout ce qui est nouveau, intéressant, bizarre, amusant ou intelligemment conçu produit sur moi un effet stimulant. La ligne de vêtements européenne DIESEL, par exemple, est des plus attrayantes: des étiquettes extraordinaires, des t-shirts imprimés à la perfection et un catalogue à l'image de la collection avec son petit côté «pas sérieux». Quant aux emballages, la meilleure carte de visite du magasin, ils sont en parfaite adéquation avec l'esprit de la marque. ■ Toute la réflexion et le soin qui entourent l'image de marque et l'emballage du produit ont naturellement tendance à se reporter sur le service à la clientèle. Le secret du designer qui parvient à créer l'emballage subtil et original reflétant fidèlement le caractère d'une ligne de produits tient tout entier dans son souci de perfection poussé jusque dans le moindre détail. Et c'est justement cette conjonction de détails qui fait que le client a plus de plaisir et de facilité à faire ses emplettes. ■ Lorsqu'un emballage me déplaît, c'est en général parce qu'il fait insulte au consommateur. Par exemple, j'ai horreur de ces emballages volumineux qui cachent des produits minuscules ou de ces montagnes de papier tout juste bonnes à remplir une poubelle. Je déteste tout autant les packagings de produits alimentaires sur lesquels la composition de l'aliment est illisible. Et j'exècre en particulier le graphisme qui recouvre inévitablement les emballages de cosmétiques ou d'articles de ménage vendus en promotion dans les grandes surfaces. De plus, ces produits sont souvent placés sous des pancartes annonçant des réductions de prix à grand renfort d'étoiles. A en croire les discounters, ce graphisme fonctionnerait comme un signal à l'adresse du consommateur. Mais dans ce contexte, le «signal» renvoie tout simplement au budget limité de la classe moyenne de la population. Selon une idée largement répandue, celui qui achète des produits en promotion part du principe que les beaux emballages sont de la poudre aux yeux dont la seule finalité est d'augmenter le prix. Il est donc primordial que les designers s'attaquent à cette idée reçue. Or, il semblerait que la seule manière que nous ayons trouvée de relever ce défi est d'apprendre aux jeunes graphistes en formation à mépriser ce langage «vulgaire» utilisé par les discounters. Une attitude qui ne sert qu'à accréditer la croyance selon laquelle le design digne de ce nom serait réservé à une élite ou que cette élite pourrait utiliser la langue des gens modestes dans un but ironique. Et pourtant, en matière d'emballage, l'argent et la qualité ne vont pas forcément de pair: les emballages dorés à inscriptions gaufrées et les flacons et autres tubes fort coûteux caractéristiques de l'industrie des cosmétiques suffisent à le prouver. Bien sûr, on trouve toujours des exceptions, comme le parfum de Jean-Paul Gaultier audacieusement vendu dans une boîte de conserve en aluminium. ■ J'imagine que les magnifiques packagings présentés dans cet ouvrage ne sont pas destinés à la consommation de masse, mais relèvent plutôt du domaine expérimental ou sont des objets promotionnels produits en série limitée. On peut sans doute en trouver quelques-uns de ce type dans certains magasins ou supermarchés assez clairvoyants pour comprendre qu'un emballage de qualité incite à la consommation en accroissant le plaisir d'acheter. Quoi qu'il en soit, il faut reconnaître que, sous la pression de la concurrence, féroce, et de la crise économique, les détaillants se montrent de plus en plus soucieux de créer une ambiance propice au shopping dans laquelle l'emballage a un rôle capital à jouer. ■ L'emballage fait partie intégrante de l'environnement. C'est l'un des rares domaines du graphisme dans lequel le produit ne peut être avalé par l'écran d'un ordinateur. En effet, l'emballage n'a rien de «virtuel». C'est un objet tridimensionnel des plus concrets. On le touche, on le sent et lorsqu'il présente un intérêt esthétique, on va jusqu'à le garder. L'emballage contribue au plaisir du shopping, l'une des rares activités, hormis le sport, qui soit encore capable d'inciter les gens à sortir de chez eux.

PAULA SCHER EST ASSOCIÉE DE PENTAGRAM DESIGN À NEW YORK. SES NOMBREUSES ACTIVITÉS EN TANT QUE DESIGNER GRAPHIQUE COMPRENNENT DES PROGRAMMES D'IDENTITÉ INSTITUTIONNELLE, DES EMBALLAGES, LA CONCEPTION GRAPHIQUE DE PUBLICATIONS ET DE MATÉRIEL DE PROMOTION. SES CRÉATIONS SONT EXPOSÉES DANS LES COLLECTIONS DU MUSEUM OF MODERN ART À NEW YORK DU MUSEUM FÜR GESTALTUNG DE ZURICH, DANS LE DENVER ART MUSEUM ET LE CENTRE GEORGES POMPIDOU À PARIS. ELLE EST MEMBRE DE L'ALLIANCE GRAPHIQUE INTERNATIONALE (AGI), DE L'AIGA ET DU COMITÉ DE RÉDACTION DE THE JOURNAL OF BRAND MANAGEMENT DE LONDRES.

PACKAGE DESIGN

PACKUNGSGESTALTUNG

FORMES D'EMBALLAGE

1

2

3

4

5 **CAHAN & ASSOCIATES** *Boisset USA*

6 **DUFFY DESIGN** *Flagstone Brewery*

(OPPOSITE) 1 **DUFFY DESIGN** *Flagstone Brewery* □ 2 **CAHAN & ASSOCIATES** *Boisset USA* □
3 **TUTSSELS** *Bass Brewers* □ 4 **DESIGN IN ACTION** *Scottish Courage*

(ABOVE) 7 **ANTISTA-FAIRCLOUGH DESIGN** *Anheuser Busch, Inc.* ☐ (BELOW) 8 **GROUP M** *Dock Street Brewing Company* ☐
(OPPOSITE) 9–12 **ANTISTA-FAIRCLOUGH DESIGN** *Anheuser Busch, Inc.* ☐ 13 **HORNALL ANDERSON DESIGN WORKS, INC.** *William & Scott Company* ☐
14 **MURRIE LIENHART RYSNER** *Kettle Moraine Beverage Co.* ☐ 15 **DIDONATO ASSOCIATES** *Goose Island Beer Company* ☐
16 **MICHAEL OSBORNE DESIGN** *St. Stan's Brewing Co.*

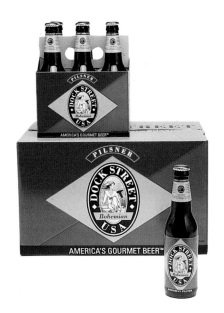

9

10

11

12

13

14

15

16

17

18

19

20

21

22

23

24

25

26 **CHRISTOPHER HADDEN DESIGN** *SHIPYARD BREWING COMPANY*

27 **DIVISION** *SAKU BREWERY, LTD.*

(OPPOSITE) 17 **TIEKEN DESIGN AND CREATIVE SERVICES** *BLACK MOUNTAIN BREWING COMPANY* □ 18 **PENTAGRAM DESIGN** *FLYING FISH BREWING COMPANY* □ 19-20 **ANTISTA-FAIRCLOUGH DESIGN** *ANHEUSER BUSCH, INC.* □ 21 **TUTSSELS** *GUINNESS BREWING UK* □ 22 **JAGER DI PAOLA KEMP DESIGN** *MAGIC HAT BREWING COMPANY* □ 23 **ANTISTA-FAIRCLOUGH DESIGN** *ANHEUSER BUSCH, INC.* □ 24 **PRIMO ANGELI** *HAL RINEY & PARTNERS* □ 25 **ANTISTA-FAIRCLOUGH DESIGN** *ANHEUSER BUSCH, INC.*

28

29

30

31

32

33 **TUCKER DESIGN** *SADDLERS CREEK WINERY*

34 **TUCKER DESIGN** *NEGOCIANTS*

(PRECEDING SPREAD) 28 **KEIZO MATSUI & ASSOCIATES** *YAGI SHIZOU-BU* □ 29–30 **DENTSU INC. KANSAI** *GEKKEIKAN* □ 31 **KEIZO MATSUI & ASSOCIATES**
YAGI SHIZOU-BU □ 32 **TCD CORPORATION** *HAKUTSURU SAKE BREWING CO., LTD. (OPPOSITE)* 35–38 **CREATION COMMUNICATION** *PROVINS VALAIS*

35

36

37

38

39 **MICHAEL OSBORNE DESIGN** *Cornerstone Cellars*

40 **DASHWOOD DESIGN** *Montana Wines*

(OPPOSITE) 41 **LYNDA WARNER, GRAPHIC DESIGN** *Chateau Xanandu* □ 42 **BLACKBURN'S LIMITED** *Cockburn Smithes & Cia Lda* □ 43 **SUPON DESIGN GROUP** *Grand Palace Food International* □ 44–46 **CATO DESIGN** *T'Gallant* □ 47 **CATO DESIGN** *Mistwood Vineyard* □ 48 **CALDEWAY DESIGN** *Tom Eddy Napa Valley Wines* □ 49 **CATO DESIGN** *T'Gallant*

41

42

43

44

45

46

47

48

49

50

51

52

53

55 **DAEDALOS** *Golan Heights Winery*

56

57

58

59

60

61

(ABOVE) 62 **THORBURN DESIGN** *Millenium* □ (BELOW) 63 **BAXMANN & HAMICKELL** *H.C. Asmussen*
(OPPOSITE) 64 **TUCKER DESIGN** *Spicers Paper*

63

65

66

64

65

66

65 **UNISON NETWORK** *Takara Shuzo Co., Ltd.* □ 66 **DESIGNERS COMPANY** *Hooghoudt BV* □
67 **DAEDALUS** *Societe Slaur* □ 68 **BLACKBURN'S LTD.** *Berry Brothers and Rudd Ltd.* □ 69 **ATELIER HAASE &**
KNELS *B. Grashoff Nachf.* □ 70 **MICHAEL PETERS LIMITED** *Courvoisier S.A. France*

36

67

68

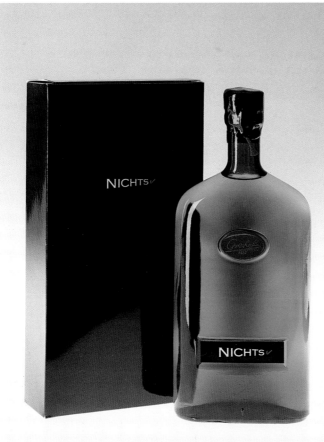

69

70

71 **DUFFY DESIGN** *Jim Beam*

72 **BLACKBURN'S LIMITED** *United Distillers*

(OPPOSITE, TOP) 73 **DAEDALUS DESIGN** *Cognac Raymond Ragnaud* □ (OPPOSITE, BOTTOM) 74 **GEORGE LACHAISE DESIGN** *Elie-Arnaud Denoix*

73

74

75 **DAEDALUS DESIGN** *France Euro Argo*   76 **DAEDALUS DESIGN** *Cognac Raymond Ragnaud*

(OPPOSITE, TOP) 77 **SOMMESE DESIGN** *Aquapenn Spring Water Co.* □ *(OPPOSITE, BOTTOM)* 78 **TANGRAM STRATEGIC DESIGN** *Gruppo Sun-Assago*

77

78

79

80

81 **DUFFY DESIGN** *The Coca-Cola Company*

(OPPOSITE, TOP) 79 **HORNALL ANDERSON DESIGN WORKS** *Talking Rain* □   *(OPPOSITE, BOTTOM)* 80 **PRIMO ANGELI** *G. Heileman Brewing Company*

82 **NESTLE BEVERAGE COMPANY** (*IN-HOUSE*)

(OPPOSITE) 83 **HOWE DESIGN** *A.G. BARR P.L.C.* □ 84–86 **DUFFY DESIGN** *THE COCA-COLA COMPANY*

83

84

85

86

87

88

89

90

91

92

93

94

95

96 **TUCKER DESIGN** *Berrivale Orchards*

(OPPOSITE) 87–89 **ANTISTA-FAIRCLOUGH DESIGN** *Royal Crown Cola* □ 90 **DASHWOOD DESIGN** *Frucor Beverages* □ 91 **GSD+M** *RC Cola/Royal Crown Company, Inc.* □ 92 **PORT MIOLLA ASSOCIATES** *Great Brands of Europe* □ 93 **FRITZSCH & MACKAT, WERBEAGENTUR** *Spreequell Mineralbrunnen GmbH* □ 94–95 **HOWE DESIGN** *A.G. Barr P.L.C.*

98 **PETER SCHMIDT STUDIOS** *EUROCOS*

(OPPOSITE) 97 **DESGRIPPES GOBÉ & ASSOCIATES** *CARTIER*

99 **PETER SCHMIDT STUDIOS** *Eurocos*

100

101

102

103

104

105 **PACKAGING CREATE INC.** *KYOSHINSHA CO., LTD.*

(OPPOSITE) 104 **DOOKIM DESIGN** *UTOO ZONE, SAMSUNG CORPORATION*

106–110 **ARAMIS INC.** *(IN-HOUSE)*

107

108

109

110

(ABOVE) 111 **DINAND DESIGN** *Tiffany & Company*
(OPPOSITE) 112-113 **PENN STATE SCHOOL OF VISUAL ARTS** *Mondo dei Sogni (student project)*

56

112

113

(THIS PAGE) 114–117 **TAMOTSU YAGI DESIGN** *Benetton Cosmetics* □ (OPPOSITE) 118–119 **TOM FOWLER, INC.** *Cheesebrough-Pond's USA Co.* □ 120 **AVANCÉ DESIGNS INC.** *Tiffany & Co.* □ 121 **VANESSA ECKSTEIN DESIGNS** *Natura* □ 122 **AVANCÉ DESIGN INC.** *Tiffany & Co.* □ 123 **HANS D. FLINK DESIGN INC.** *Fabergé Co.*

115

116

117

118

119

120

121

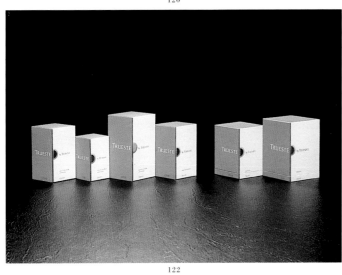

122

123

1957

The 1957 Corvettes were sculptured works of automotive art. They were devoid of the unnecessary accessories and frills so common to the era. The introduction of two options, fuel injection and a four speed transmission, made the 1957 model more desirable. The 283 fi V8 showed horse on the racing circuit and capped off the season with Dick Thompson's victory in the Sports Car Club of America championship.

1967

The 1967 Corvette is thought by many to be the most refined Sting Ray of all. Corvette enthusiasts know it to be a very rare and highly prized investment. How many collectors haven't heard about the 435 hp 1967 convertible selling for Ferrari prices? The L88 racing version was clocked at 171 mph on the In Men's Mulsanne straight, which was as much as 20 mph faster than the fastest Ferrari entered.

1969

1969 marked the second year of the longest running body style in Corvette history. Born of the Mako Shark show car, the body would not be changed again until 1968. The side exhaust system added a beauty no one could deny and was unique to 1969. This was the second year of the ever popular "T" top. However, it is not until the 435 hp 1967 CI engine revs up and roars away, that its awesome beauty is revealed.

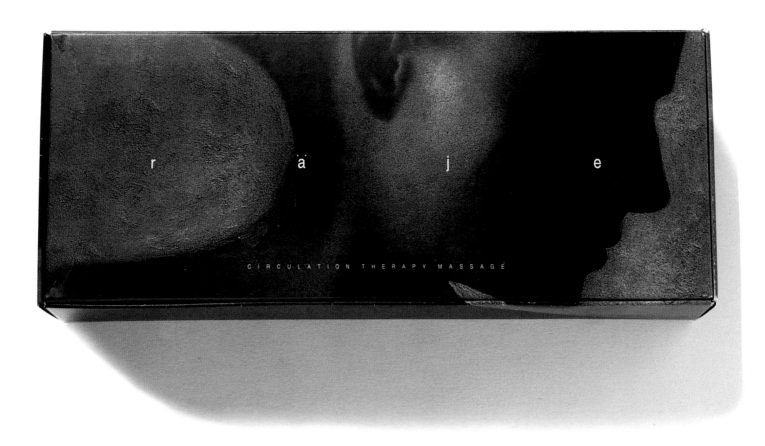

(ABOVE) 127 **ADRIAN PULFER DESIGN** *RAJE* □ (BELOW) 128 **MARY KAY INC.** (*IN-HOUSE*)
(OPPOSITE) 129 **AVON PRODUCTS CO., LTD.** (*IN-HOUSE*)

128

131

130 **LONCZAK-FILIZ DESIGN** *COSMAIR* □ 131 **DESIGN IN ACTION** *COTY*
(OPPOSITE) 132–134 **DFS GROUP, LTD.** *DFS MERCHANDISING, LTD.*

132

133

134

(OPPOSITE) 135 **LEWIS MOBERLY** *Next PLC.* □ (ABOVE, LEFT) 136 **ANTISTA-FAIRCLOUGH DESIGN** *Mont Source, Inc.*
(ABOVE, RIGHT) 137 **PHARMA PERFORMANCE GMBH** *Simons GmbH*

138 **ANTISTA-FAIRCLOUGH DESIGN** *Mont Source, Inc.*

139–140 **DESGRIPPES GOBÉ & ASSOCIATES** *VICTORIA'S SECRET*

 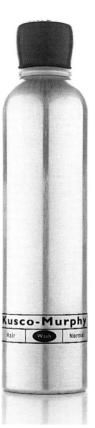

(ABOVE) 141–142 **FABIO ONGARATO DESIGN** *Kusco-Murphy Salon Pty. Ltd.*

(OPPOSITE) 143–144 **CLAUS KOCH CORPORATE COMMUNICATIONS** *Wella AG* □ 145–146 **ANGELO SGANZERLA** *L'Erbolario* □ 147 **DESIGN IN ACTION** *Coty* □ 148 **DESGRIPPES GOBÉ & ASSOCIATES** *Boucheron* □ 149–150 **DESGRIPPES GOBÉ & ASSOCIATES** *Victoria's Secret*

143

144

145

146

147

148

149

150

(THIS SPREAD) 151–157 **FITCH INC.** HUSH PUPPIES COMPANY

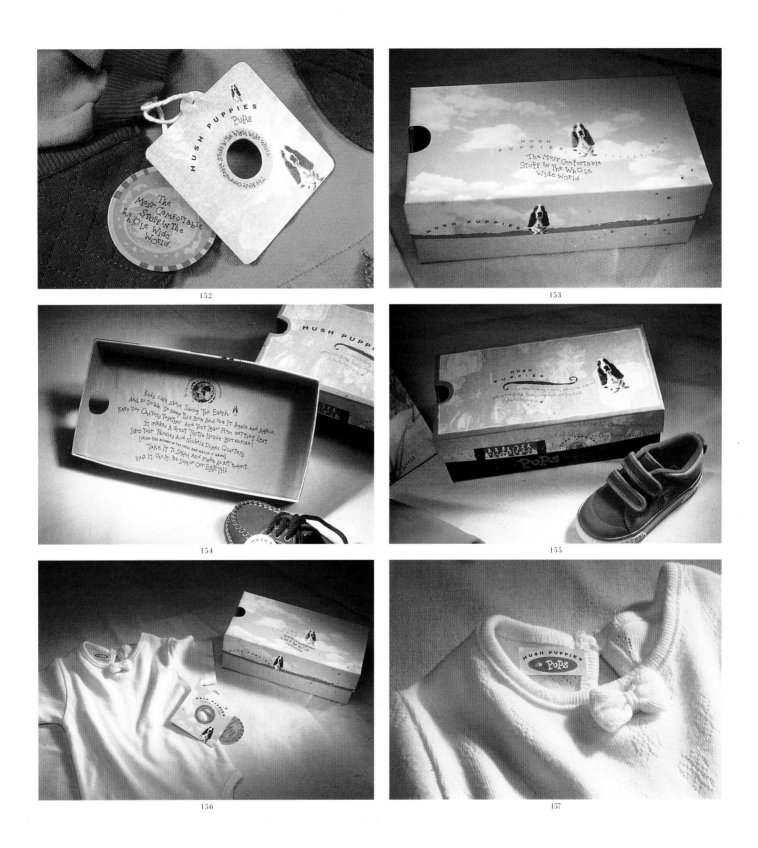

152

153

154

155

156

157

158 **PARHAM SANTANA, INC.** *Via International Group*

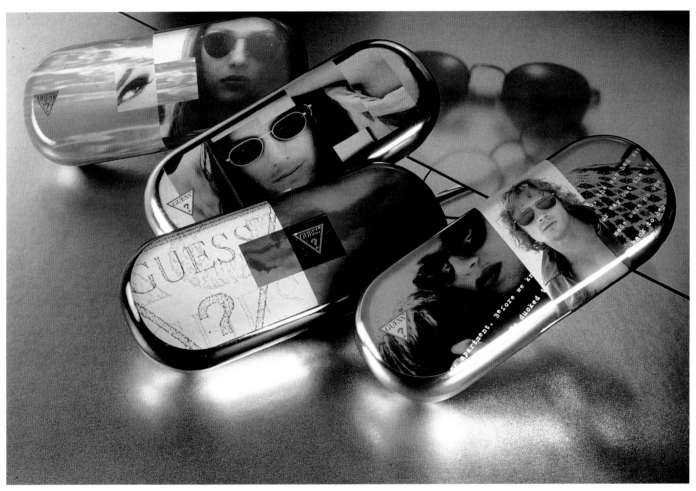

159 **PARHAM SANTANA, INC.** *Via International Group*

160 **SAZABY GRAPHIC DESIGN** *SAZABY INC.*

(*OPPOSITE*) 161 **ALAN CHAN DESIGN COMPANY** *SWANK INSPIRATION*

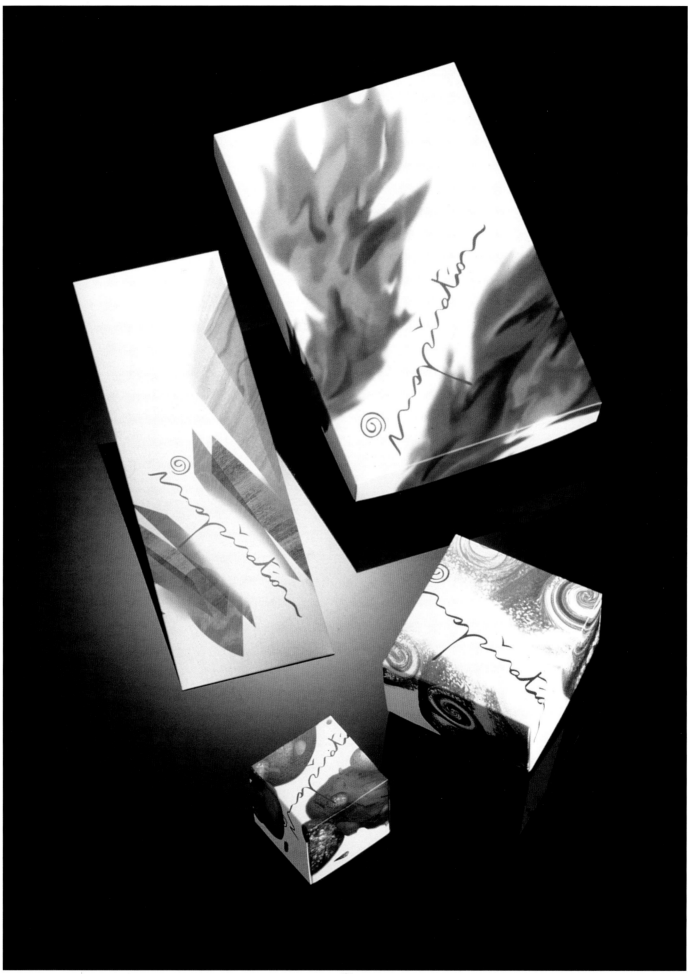

161

162

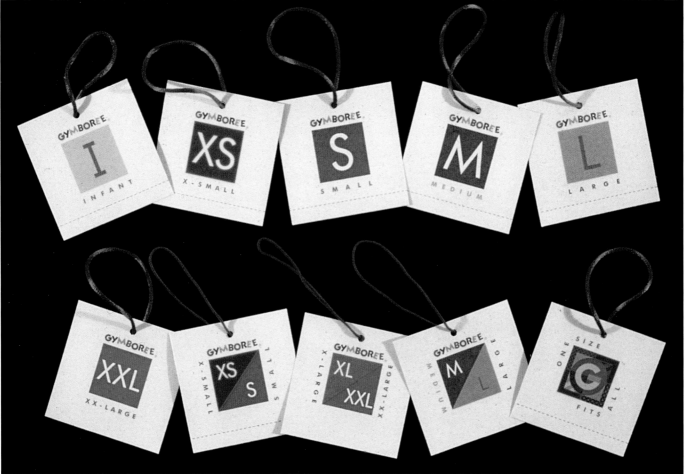

163

165

166

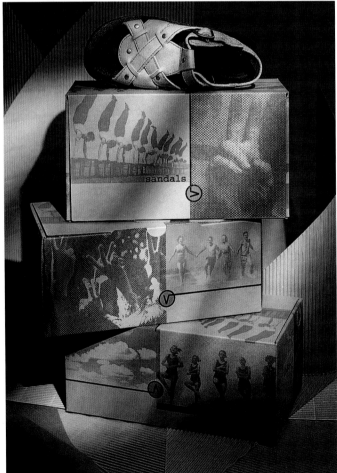

167

168

169 **LINDA FOUNTAIN DESIGN** *STRUNG OUT DESIGN*

171 **DFS GROUP, LTD.** *DFS Merchandising, Ltd.*

172 **PACKAGE LAND COMPANY LTD.** *Arab Coffee Company Ltd.*

173

174

175

176

177 **WEISS, WHITTEN, STAGLIANO** *BUCKS COUNTY COFFEE*

178 **WEISS, WHITTEN, STAGLIANO** *BUCKS COUNTY COFFEE*

(OPPOSITE) 173 **ELTON WARD DESIGN** *(IN-HOUSE)* □ 174–175 **CERADINI DESIGN** *COMPASS FOODS* □ 176 **HORNALL ANDERSON DESIGN WORKS, INC.** *STARBUCKS COFFEE COMPANY*

179

180

181

182

(THIS SPREAD) 179–183 **DUFFY DESIGN** The Coca-Cola Company

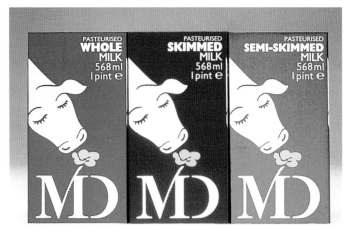

184

185

186

187

188

189 **MIRES DESIGN INC.** *Ken C. Smith Advertising*

190 **WERNER DESIGN WERKS** *Target Stores*

191 **KAN TAIN-KEUNG DESIGN & ASSOCIATES** *EFFEM FOODS INC.*

192

193

194

195

196

197

192-193 **ANGELO SGANZERLA** *Andrea Stainer* □ 194-195 **DFS GROUP, LTD.** *DFS Merchandising, Ltd.* □
196-197 **SACKETT DESIGN ASSOCIATES** *DFS Group Limited*

198 **MURRIE LIENHART RYSNER & ASSOCIATES** *DAYTON-HUDSON*

199 **ATELIER ZONE** *OS3, ORGANISATION*

(OPPOSITE) 200–203 **TOWER SHOP** *NDC GRAPHICS*

200

201

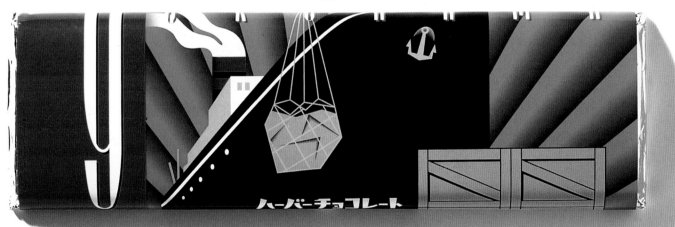

202

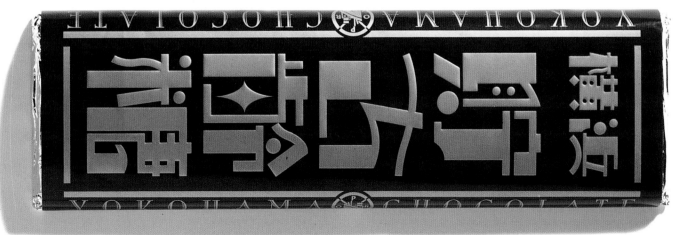

203

211–212 **ALAN CHAN DESIGN COMPANY** *MR CHAN TEA ROOM*

(OPPOSITE) 213–217 **SANDSTROM DESIGN** *TAZO TEA COMPANY*

213

214

215

216

217

218 **TASSILO VON GOLMAN DESIGN** *CLUB ENGLISH TEA*

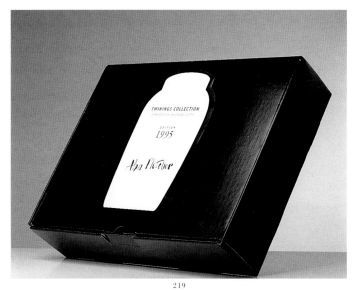

219

220

221

222

223

224

219 **TASSILO VON GROLMAN DESIGN** *CLUB ENGLISH TEA* □ 220 **TASSILO VON GROLMAN DESIGN** *TWININGS ICED TEA* □
221 **DFS GROUP, LTD.** *DFS MERCHANDISING, LTD.* □ 222 **VINEYARD DESIGN LIMITED** *J SAINSBURY PLC* □ 223 **PACKAGE LAND CO., LTD.** *TAMAYA CO., LTD.* □
224 **KAN TAIN-KEUNG DESIGN & ASSOCIATES** *UNILEVER HONG KONG LTD.*

225 **TUCKER DESIGN** *Lactos*

(OPPOSITE) 226 **CERADINI DESIGN** *The Great Atlantic & Pacific Tea Company*

227-228 **FÖLSER & SHERNHUBER** *Anton Bürstunger Landfleischerei*

227

228

(OPPOSITE) 229-230 **BOLT, KOCH & COMPANY** *Volg Konsumwaren AG*

229

230

231 **MILLFORD-VAN DEN BERG DESIGN** *Albert Heijn B.V.*

(OPPOSITE) 232 **THOMPSON DESIGN GROUP** *Pacific Grain Products* □ 233 **PRIMO ANGELI** *Arrowhead Mills* □
234 **LANDOR ASSOCIATES** *J.M. Smuckers* □ 235 **LANDOR ASSOCIATES** *H.E.B.* □ 236 **ANGELO SGANZERLA** *Andrea Stainer* □ 237 **CERADINI DESIGN**
The Great Atlantic & Pacific Tea Company □ 238 **PRIMO ANGELI** *Del Monte* □ 239 **MONNENS-ADDIS DESIGN** *Bell-Carter Foods*

232

233

234

235

236

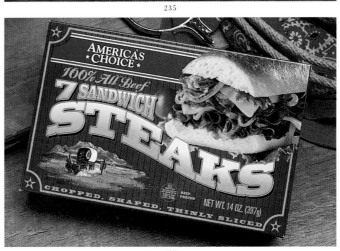

237

238

239

240 **PACKAGE & LOGO DESIGN** *APPEL & FRENZEL GMBH*

241 **ATELIER HAASE & KNELS** *B. GRASHOFF NACHF*

242

243

(THIS SPREAD) 242–245 **PENTAGRAM DESIGN** *COLUMBUS SALAME*

244

245

246

247

(ABOVE) 246 **ANTISTA-FAIRCLOUGH DESIGN** *Wolfgang Puck Food Company* ▢ 247 **BRITTON DESIGN** *Viansa Winery* ▢ *(OPPOSITE)* 248–249 **SLAUGHTER-HANSON** *Bottega Restaurant/Cafe* ▢ 250 **FÖLSER + SCHERNHUBER** *Sonnentor Kräuterhandelsges, GmbH* ▢ 251 **JACOBSON ROST** *Schreier Malting Co.*

110

248

249

250

251

252

253

254

(THIS SPREAD) 252-254 **DFS GROUP, LTD.** DFS MERCHANDISING, LTD

255 **PENTAGRAM DESIGN** *DOUBLETREE HOTELS CORPORATION*

(OPPOSITE) 256 **HORNALL ANDERSON DESIGN WORKS** *CONTINENTAL MILLS*

257

258

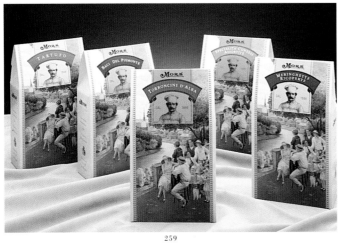

259

260

261

262

263

264

265 **ROTTKE WERBUNG** *IRISH DAIRY BOARD*

266

267

268

269

266–269 **SUPON DESIGN GROUP INC.** *FRESH MARKET* □ *(OPPOSITE)* 270 **ANGELO SGANZERLA** *L'ERBAOLARIO* □ 271 **GEORGE LACHAISE DESIGN** *ELIE-ARNAUD DENOIX* □ 272 **CHRISTENSEN/LUND** *REGAL MOLLE, STARBBURET* □ 273 **CHRISTENSEN/LUND** *NORSKE MEIERIER* □ 274 **GAYLORD GRAPHICS** *CALIFORNIA TOMATO PICKERS* □ 275 **BOLT, KOCH & CO.** *VOLG KONSUMWAREN AG* □ 276 **DAMORE-JOHANN DESIGN** *PHILCHICK, INC.* □ 277 **BOLT, KOCH & CO.** *VOLG KONSUMWAREN AG*

270

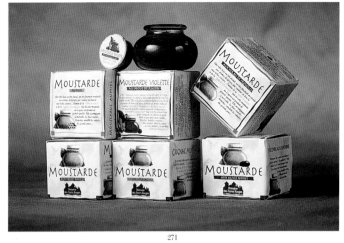

271

272

273

274

275

276

277

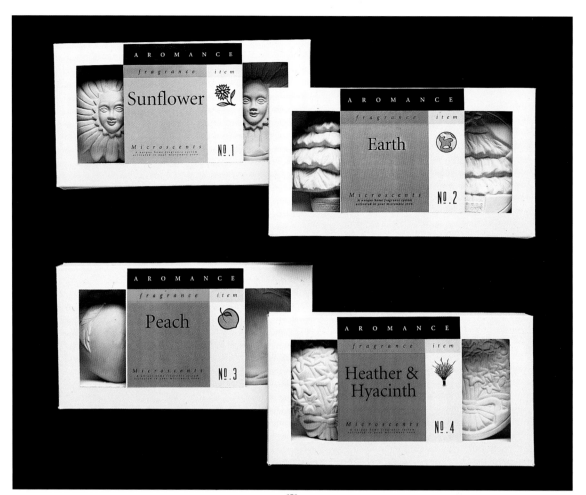

278

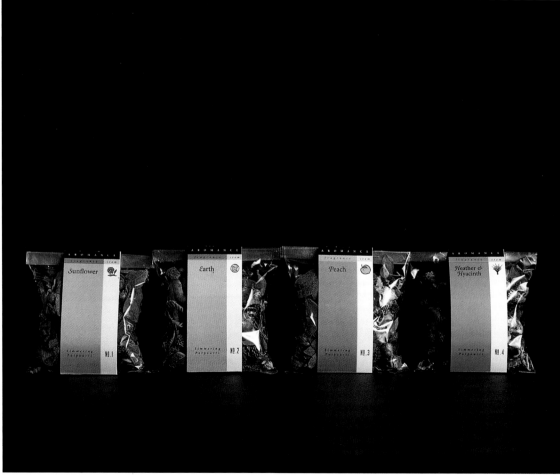

279

280

281

121

282 **TUTSSELS** *Boots*

(PRECEDING SPREAD) 278–281 **LIDJI DESIGN OFFICE** *AROMANCE HOME FRAGRANCES*

283 **WERBEATELIER FICK WERBEAGENTUR GMBH** *Rosenthal AG*

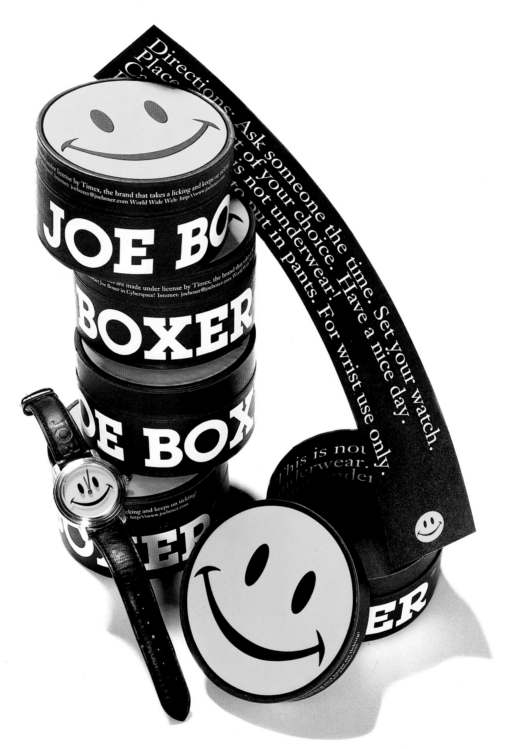

284 **SMART DESIGN** *TIMEX, JOE BOXER*

(OPPOSITE) 285 **PHILLIPS DESIGN GROUP** *ATLANTIC TECHNOLOGY* ☐ 286 **MICHAEL OSBORNE DESIGN** *EMPIRE BEROL USA*

285

286

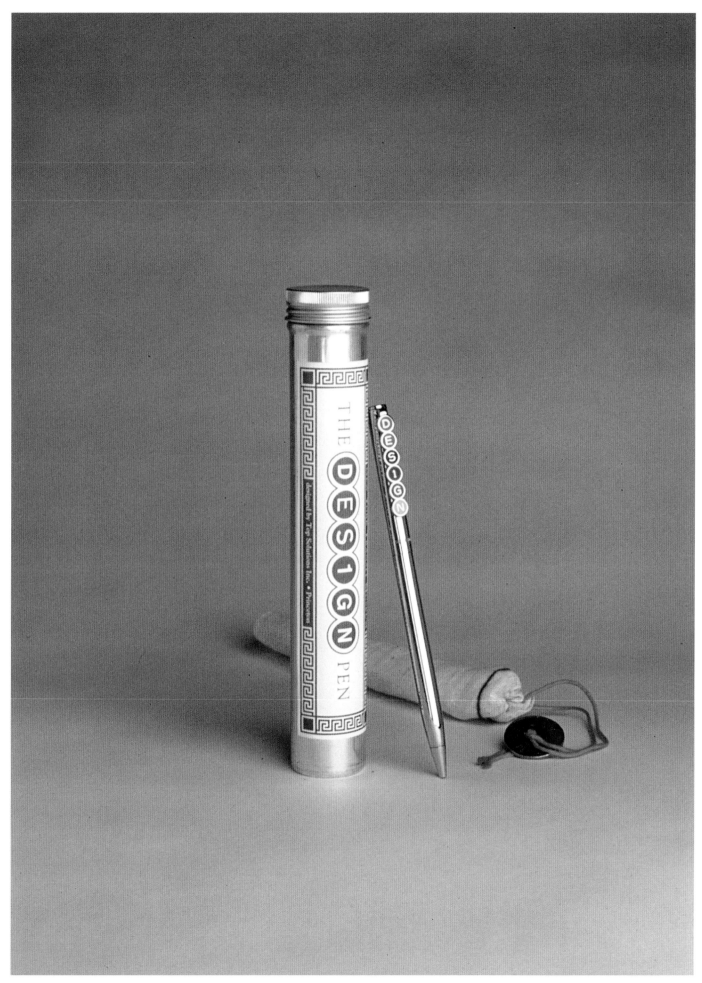

288 **TOP SPIN DESIGN** *Metropolitan Transit Authority*

289

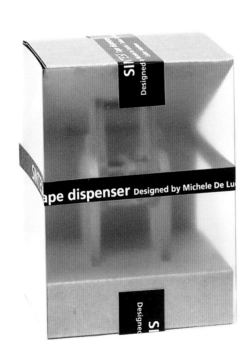
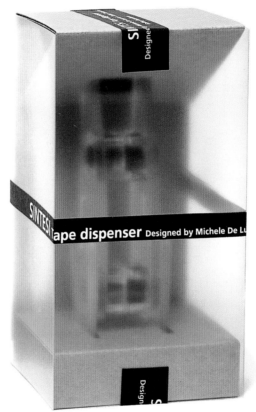

290

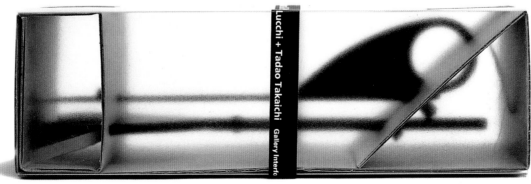

291

292

(THIS SPREAD) 289–292 **JUN SATO DESIGN, INC.** GALLERY INTERFORM

293 **LOVE PACKAGING GROUP** *THE HAYES COMPANY INC.*

(OPPOSITE) 294 **FITCH INC.** *DIGITAL EQUIPMENT CORPORATION*

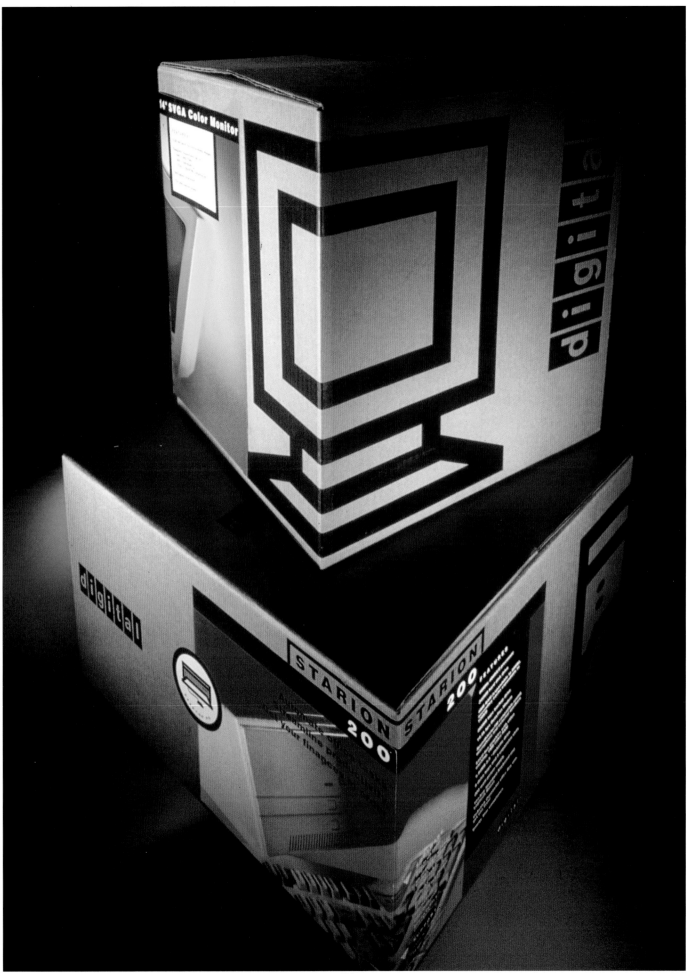

294

295

296

297

298 **SANDSTROM DESIGN** *LEUPOLD*

(*OPPOSITE*) 295 **CHRISTENSEN/LUND** *JOTUN A/S* □ 296 **HARCUS DESIGN** *WATTYL AUSTRALIA* □ 297 **CHRISTENSEN/LUND** *JOTUN A/S*

299 **HORNALL ANDERSON DESIGN WORKS, INC.** *OXO International*

(OPPOSITE) 300 **HORNALL ANDERSON DESIGN WORKS, INC.** *OXO International* □ 301 **DGWB ADVERTISING** *Qualcomm Inc.* □ 302 **DUFFY DESIGN** *Yakima* □
303 **SIEGER DESIGN CONSULTING GMBH** *RC Ritzenhoff Cristal GmbH*

300

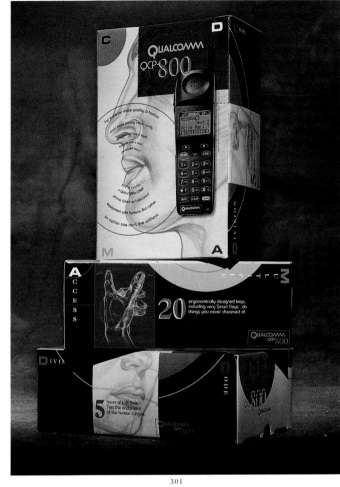

301

302

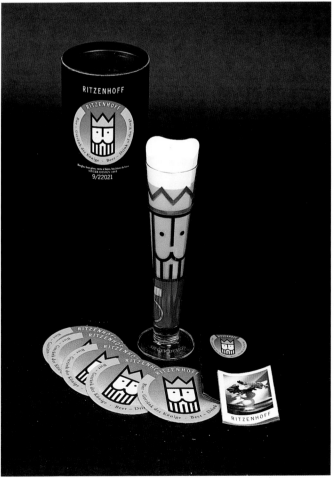

303

304

305

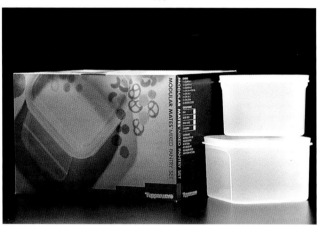

306

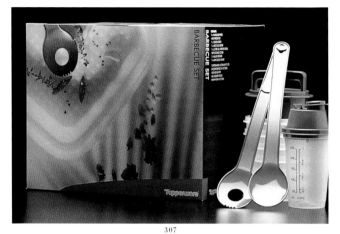

307

308

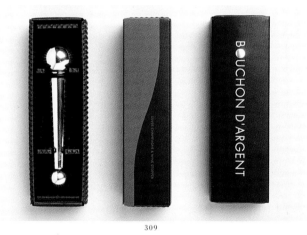

309

310

311

312–313 **TASSILO VON GROLMAN DESIGN** *Frankfurte grüne Sauce*

314

315 **HUNDRED DESIGN, INC.** *(In-House)*

(OPPOSITE) 314 **TUTSSELS** *Philips Lighting*

316

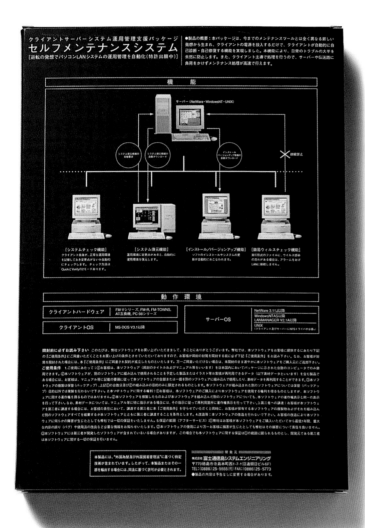

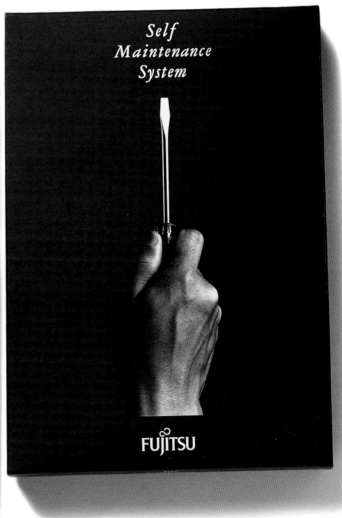

317-318 **DAIKO ADVERTISING, INC.** *FUJITSU TOKUSHIMA SYSTEMS ENGINEERING LTD.*

(OPPOSITE) 316 **PARHAM SANTANA, INC.** *M.H. SEGAN & COMPANY*

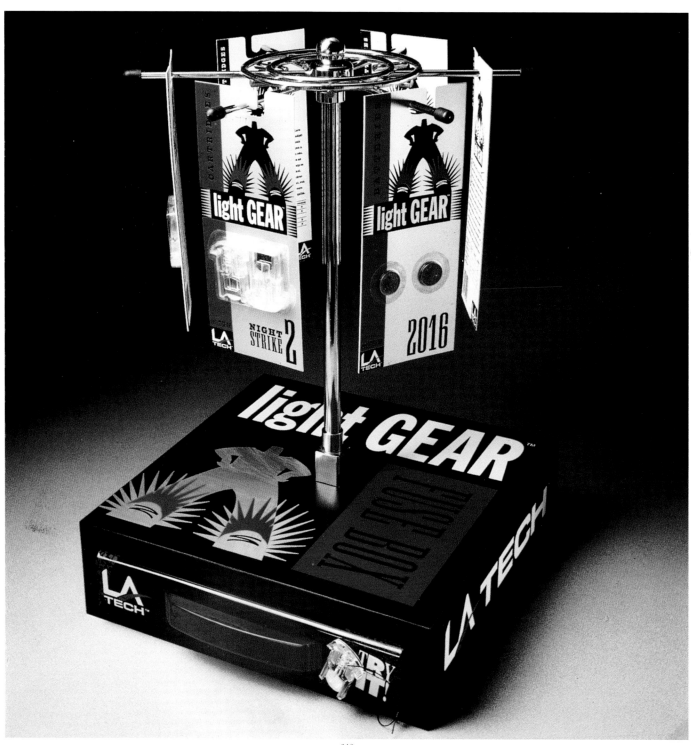

319

319 **MIRES DESIGN INC.** *LA Gear* □ (OPPOSITE)320 **FITCH INC.** *Digital Equipment Corporation* □ 321 **PACKAGE LAND CO.,**
LTD. *Saiwai Co., Ltd.* □ 322 **FRANKFURT BALKIND PARTNERS** *Pantone, Inc.*

320

321

322

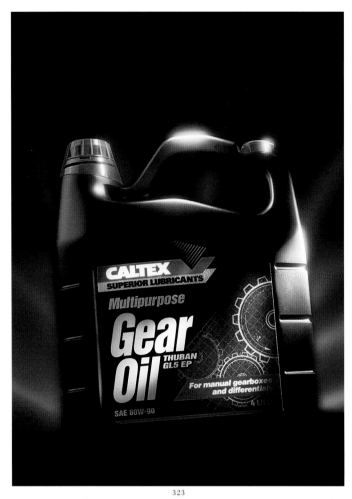

323

324

325

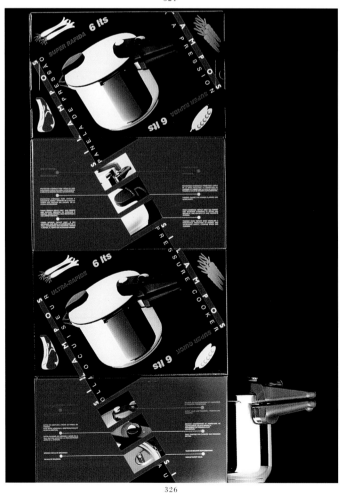

326

327 **GADESI** *SILAMPOS*

(OPPOSITE) 323 **ELTON WARD DESIGN** *CALTEX OILS AUSTRALIA* □ 324 **PACKAGE LAND CO., LTD.** *EIKOSHA CO., LTD.* □
325 **LIPPA PEARCE DESIGN** *HALFORD'S LIMITED* □ 326 **GADESI** *SILAMPOS*

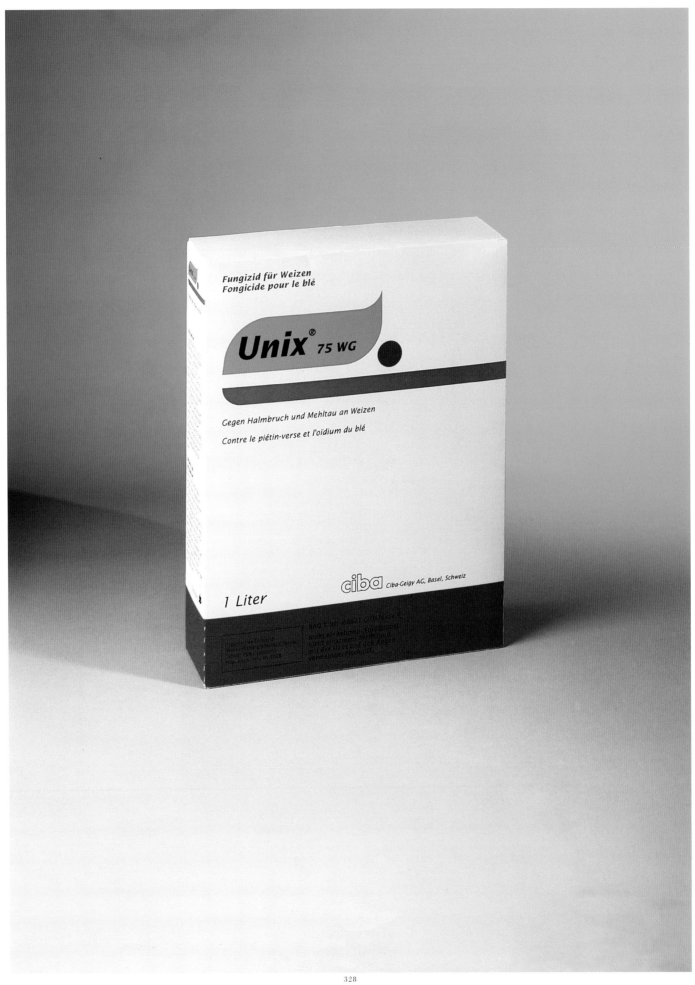

329

330

331 **PELLONE & MEANS** *MANOSTAT CORPORATION*

(PRECEDING SPREAD) 328 **SCHAFFNER & CONZELMANN AG** *CIBA AGRON SCHWEIZ* □ 329 **CORNERSTONE** *CHEBBY FOODS* □ 330 **GARDEN STUDIO** *MILFORD-VAND DEN BERG DESIGN/SWEETLIFE B.V.* □ *(OPPOSITE)* 332–333 **PHARMA PERFORMANCE GMBH** *MUNDIPHARMA GMBH*

332

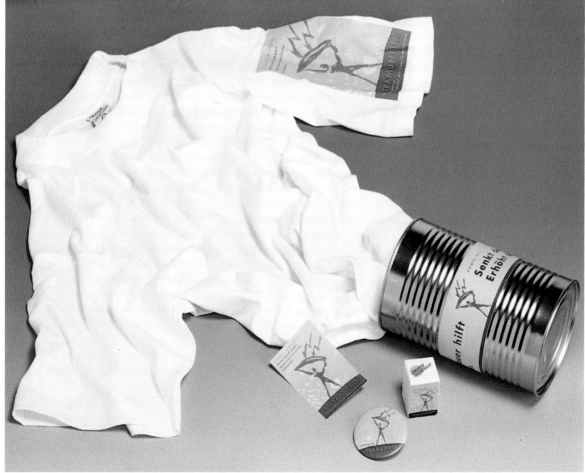

333

334

335

336

337

338 **JENSEN DESIGN ASSOCIATES** *CANON COMPUTER SYSTEMS, INC.*

(OPPOSITE) 334 **AKAGI REMINGTON** *OPCODE MUSIC SYSTEMS* □ 335 **MADDOCKS AND CO.** *SONY COMPUTER ENTERTAINMENT* □
336–337 **THE DESIGN OFFICE OF WONG & YEO** *DIGITAL PICTURES*

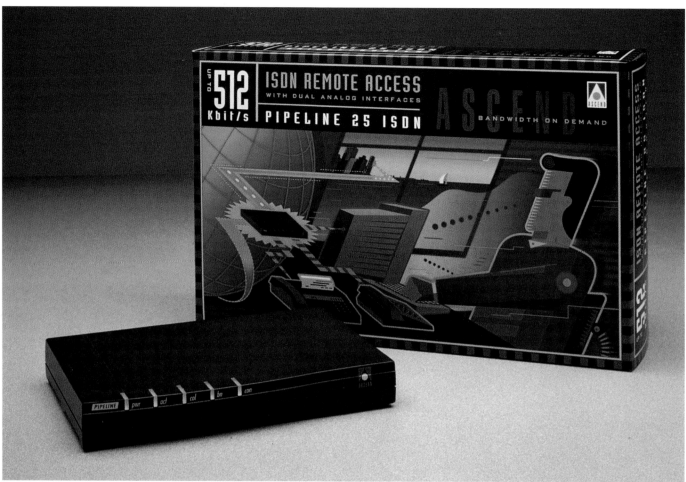

339

340

341

342

343

344

345

346

348

349

350

351

353

354

355

356

357

358

359

360

361

362

363

364

365

366

367

368

369 **DUFFY DESIGN** *Ergodyne*

370 **PENTAGRAM DESIGN, INC.** *Computer Curriculum Corp.*

371-373 **GALLERY INTERFORM** *ART AGAINST AIDS PROJECT*

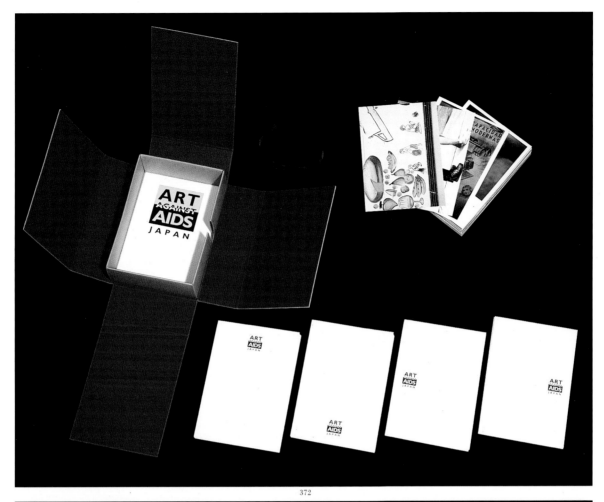

372

373

374

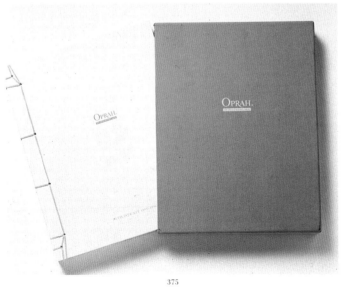

375

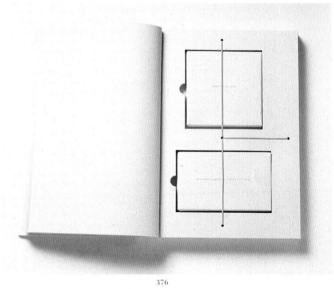

376

374 **THE PUSHPIN GROUP** *HARRY N. ABRAMS* □ 375–376 **VSA PARTNERS, INC.** *HARPO PRODUCTIONS, INC.* □ *(OPPOSITE)* 377–378 **FÖLSER + SCHERNHUBER** *(IN-HOUSE)*

379

380

379 **HANSON DESIGN** *Wattu Wear, Inc.* □ 380 **DUFFY DESIGN** *Yakima (OPPOSITE)* 381 **HORNALL ANDERSON DESIGN WORKS, INC.** *Smith Sport Optics* □
382 **FRANKFURT BALKIND PARTNERS** *Pantone, Inc.*

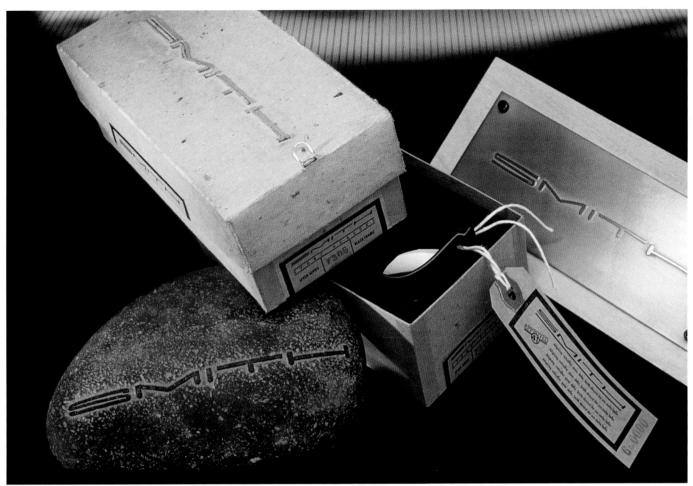

381

382

383

384

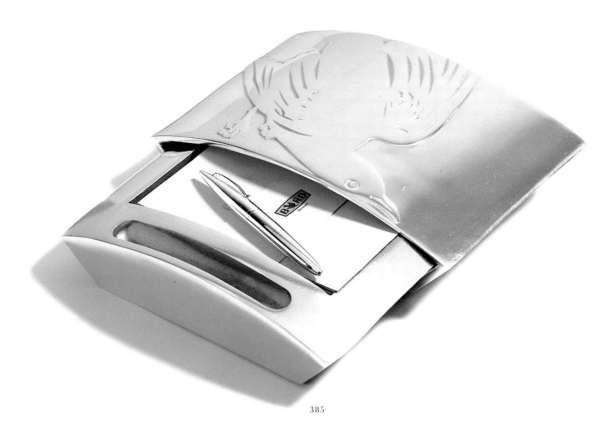

385

386

(OPPOSITE) 383–384 **SHR PERCEPTUAL MANAGEMENT** *Land Rover of North America* (ABOVE) 385–386 **BRD DESIGN** *(In-House)*

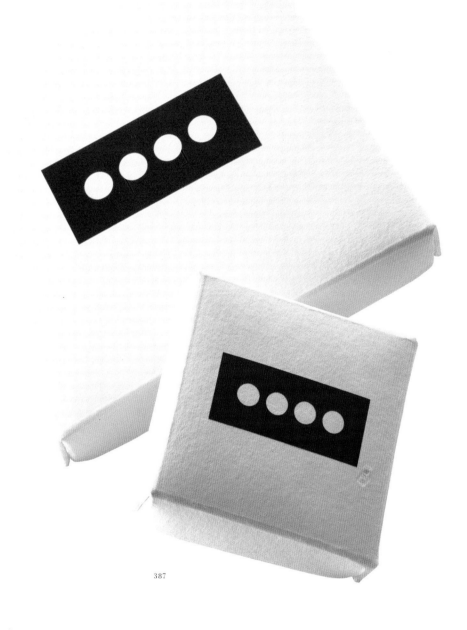

387

387 **PACKAGING CREATE INC.** *New Oji Paper Co. Ltd.*
(OPPOSITE) 388–389 **NORTHLICH STOLLEY LAWARRE DESIGN GROUP** *Mead Communication Papers*

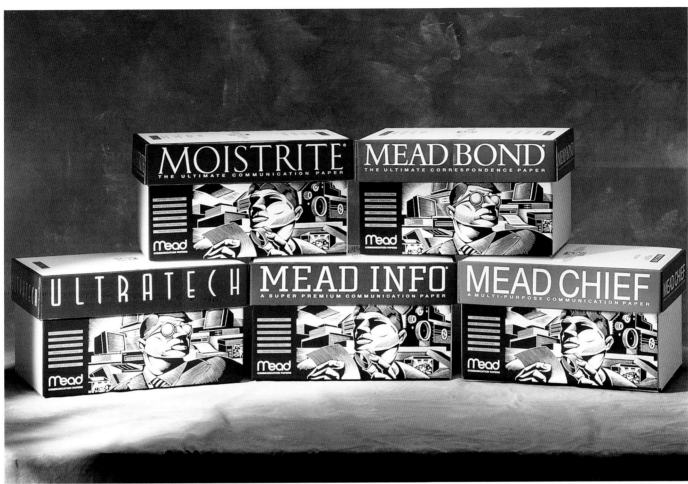

388

389

171

391

392

393

394

391–394 **DESIGNERS COMPANY** *Hooghoudt Distillers B.V.*

395

396

397

398

399 **GRETEMAN GROUP** *Hays Co.*

400 **TUTSSELS** *(IN-HOUSE)*

401

402

403

404

405

406

407

408

409

410

411

412

413

414

(ABOVE) 414 **ATELIER HAASE & KNELS** BOUTIQUE EVELYN (OPPOSITE) 415–416 **PACKAGE LAND CO., INC** (IN-HOUSE)

415

416

417 **PACKAGING CREATE INC.** *New Oji Paper Co., Ltd.*

418 **DOOKIM DESIGN** *LANEIGE, PACIFIC GROUP*

419 **MUI & GRAY** *AKI S.P.A.*

(OPPOSITE) 420 **ALAN CHAN DESIGN COMPANY** *KOSTA BODA*

420

(THIS PAGE) 421 **ARAMIS, INC.** *(IN-HOUSE)*□ *(OPPOSITE, TOP)* 422 **PENTAGRAM DESIGN** *GIANFRANCO LOTTI* □
(OPPOSITE, BOTTOM) 423 **DFS GROUP, LTD.** *DFS MERCHANDISING, LTD.*

422

423

424 **SASART GRAFIK & DESIGN** *N.A.P. Production*

(OPPOSITE) 425 **HORNALL ANDERSON DESIGN WORKS, INC.** *Jamba Juice* □ 426 **ANGELO SGANZERLA** *L'Erbaolario* □ 427 **GENSLER & ASSOCIATES**
Stephanie Leigh's □ 428 **ALAN CHAN DESIGN COMPANY** *Hong Kong Tourist Association* □ 429 **ORBIT CITY ART COMPANY** *Hanna-Barbera* □
430 **HORNALL ANDERSON DESIGN WORKS** *University Village* □ 431 **SUPON DESIGN GROUP** *Museum of Junk* □ 432 **AREA STRATEGIC DESIGN**
Instituto Nazionale Confederale di Asistenza (INCA) □ 433 **AREA STRATEGIC DESIGN** *Farmindustria*

425

426

427

428

429

430

431

432

433

434

435

436

(OPPOSITE) 434–435 **AREA STRATEGIC DESIGN** BERNI □ (ABOVE) 436 **DESGRIPPES GOBÉ & ASSOCIATES** ANN TAYLOR LOFT

437

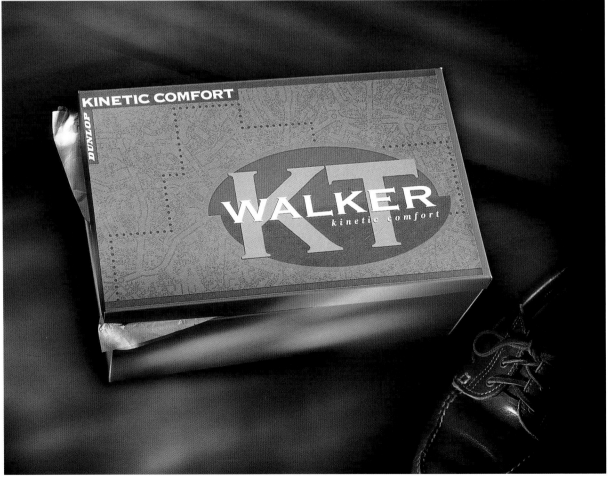

438

439

440

441 **MIRES DESIGN INC.** *VOIT SPORTS, INC.*

443

444

445

446

447

(OPPOSITE) 443 **MADDOCKS & COMPANY** *BELL HELMETS* ☐ 444-445 **LEWIS MOBERLY** *HALFORDS LTD* ☐
(ABOVE) 446-447 **STANASZEK GOODWIN DESIGN PARTNERSHIP** *BETA SPORTS, INC.*

197

448–449 **SPARTA DESIGNEREI** *Adidas Germany*

(OPPOSITE) 450 **CALDERA DESIGN** *Deckers Outdoor Corporation* □ 451 **MIRES DESIGN INC.** *Voit Sports, Inc.* □ 452 **MIRES DESIGN INC.** *Ektelon* □ 453 **PLANET DESIGN COMPANY** *Graber USA* □ 454 **JAGER DI PAOLA KEMP DESIGN** *Burton Snowboards* □ 455 **DUFFY DESIGN** *Duke Design Inc.* □ 456–458 **SPARTA DESIGNEREI** *Adidas Germany*

450

451

452

453

454

455

456

457

458

459-460 **ATELIER HAASE & KNELS CO.** *STANWELL VERTRIEBS GMBH*

460

461

461–463 **CORNERSTONE** *Moonlight Tobacco Co.*

462

463

464

465

204

466 **INTERNATIONAL IMAGING INC.** *R.J. Reynolds International B.V.*

(OPPOSITE) 464 **INTERNATIONAL IMAGING INC.** *R.J. Reynolds International B.V.* □ 465 **CORNERSTONE** *Moonlight Tobacco Co.*

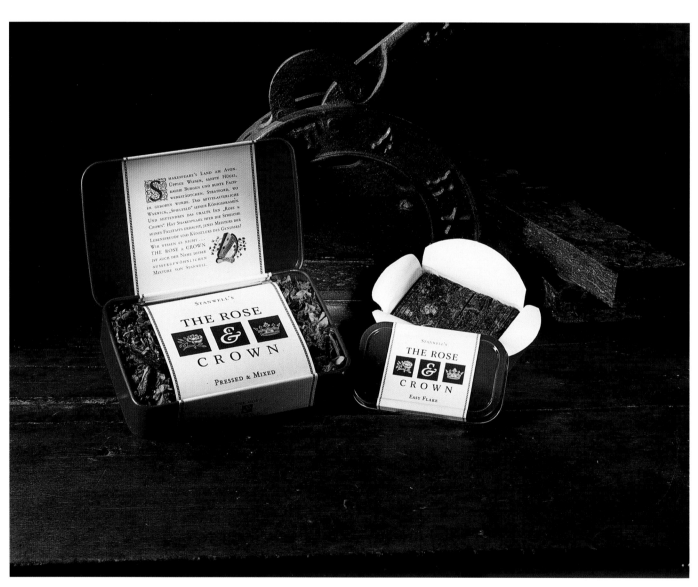

467 **ATELIER HAASE & KNELS** *STANWELL VERTRIEBS GMBH*

(OPPOSITE) 468–471 **ATELIER HAASE & KNELS** *STANWELL VERTRIEBS GMBH*

468

469

470

471

472 **ATELIER HAASE & KNELS** *STANWELL VERTRIEBS GMBH*

CAPTIONS AND INDICES

LEGENDEN UND KÜNSTLERANGABEN

LÉGENDES ET ARTISTES

PAGE 18, *image 1* ART DIRECTOR: *Dan Olson* DESIGNERS: *Dan Olson, Todd Bartz, Eden Fahlen* PHOTOGRAPHER: *Lev Tushaus* AGENCY: *Duffy Design* CLIENT: *Flagstone Brewery* COUNTRY: *USA* ■ The assignment was to brand, package, and promote a Southeastern regional specialty. ● *Der Auftrag erstreckte sich auf die Gestaltung von Logo, Verpackung und Promotionsmaterial für ein Bier, das als Spezialität der Region vermarktet wird.* ▲ *Ce contrat comprenait la conception du logo, de l'emballage et du matériel de promotion pour une bière lancée sur le marché en tant que spécialité régionale.*

PAGES 18, 19; *images 2, 5* ART DIRECTOR: *Bill Cahan* DESIGNER: *Kevin Roberson* Product PHOTOGRAPHER: *Tony Stromberg* AGENCY: *Cahan & Associates* CLIENT: *Boisset USA* COUNTRY: *USA* TYPEFACE: *Handlettering* ■ This packaging was designed to convey the character of the Prohibition era; many inmates of the Alcatraz penitentiary were arrested for bootlegging alcohol. ● *Dieses Bier wurde 'Alcatraz' genannt, weil viele Strafgefangene der berüchtigten Strafanstalt beim Schwarzbrennen erwischt wurden. Die Verpackung sollte dementsprechend an die Prohibitionszeit in den USA erinnern.* ▲ *Le nom de cette bière s'inspire directement du célèbre établissement pénitencier d'Alcatraz. De nombreux prisonniers se sont retrouvés derrière les barreaux parce qu'ils distillaient clandestinement de l'alcool. L'emballage rappelle la période de prohibition aux Etats-Unis.*

PAGE 18, *image 3* ART DIRECTOR: *Glenn Tutssel* DESIGNER: *Garrick Hamm* AGENCY: *Tutssels* CLIENT· *Bass Brewers* COUNTRY: *England* TYPEFACE: *Gill and Universe* ■ The silver labels and ice skating lines of this design reflect the product identity of this new "ice beer." ● *Die silbrigen Etiketten und die in der Gestaltung verwendeten Spuren von Schlittschuhen sollen die Positionierung des Biers als «Eis-Bier» unterstützen.* ▲ *Les étiquettes argentées et les traces de patins utilisées pour le design graphique doivent positionner cette boisson en tant que bière «glacée».*

PAGE 18, *image 4* ART DIRECTOR: *Mark Chittenden* DESIGNER: *Elaine Barbook* TYPOGRAPHY: *Design in Action* AGENCY: *Design in Action* CLIENT: *Scottish Courage* COUNTRY: *UK* ■ A new brand targeting women for a premium, lager-based product. ● *Neueinführung einer Biermarke, die sich an die Frau als Konsumentin von hochklassigem Lagerbier wendet.* ▲ *L'objectif était de créer une nouvelle marque pour cette bière blonde de qualité supérieure qui s'adresse aux femmes.*

PAGE 19, *image 6* ART DIRECTOR: *Dan Olson* DESIGNERS: *Dan Olson, Todd Bartz, Eden Fahler* PHOTOGRAPHER: *Lev Tushaus* AGENCY: *Duffy Design* CLIENT: *Flagstone Brewery* COUNTRY: *USA* ■ The assignment was to brand, package, and promote a Southeastern regional specialty using an 1860s southern brewing recipe. ● *Hier ging es um die Marke, Verpackung und Promotionsmaterial für ein Spezialbier, das nach einem Rezept von 1860 aus dem Südosten der USA gebraut wird.* ▲ *Le contrat comprenait la marque, l'emballage et la promotion d'une bière du Sud-Est des Etats-Unis, une spécialité régionale brassée selon une recette datant de 1860.*

PAGE 20, *image 7* ART DIRECTORS: *John Marota, Thomas Fairclough, Tom Antista* DESIGNER: *Thomas Fairclough* PHOTOGRAPHY: *Michael West Photography* AGENCY: *Antista Fairclough Design* CLIENT: *Anheuser Busch Inc.* COUNTRY: *USA* ■ The packaging was developed from 1800s resource material from the Anheuser Busch archives. ● *Diese Verpackung wurde auf der Basis von Archivmaterial der Brauerei aus der Zeit um 1800 entwickelt.* ▲ *Cet emballage a été créé sur la base des archives de la brasserie qui remontent aux années 1800.*

PAGE 20, *image 8* ART DIRECTOR: *Thomas Sakol* DESIGNER: *Anita Bassie* ARTISTS/ILLUSTRATORS: *various* AGENCY: *Group M* CLIENT: *Dock Street Brewing Company* PRINTER: *various* TYPEFACE: *Bodoni* ■ The goal was to make the product stand out from those of other microbreweries and to reposition the beer among premium European beers. The three primary colors represent the three varieties of beer produced by the client. ● *Aufgabe dieser Verpackung war die Aufwertung der Marke bei den Zielgruppen der kleinen Brauereien sowie bei den Zielgruppen der europäischen Biere. Die drei Farben stehen für die drei verschiedenen Biersorten.* ▲ *Cet emballage avait pour but de marquer le produit de ceux des autres petites brasseries et de repositionner la bière du client parmi les principales bières européennes de qualité supérieure. Les trois couleurs primaires utilisées représentent les trois bières que le client produit.*

PAGE 21, *image 9* ART DIRECTORS: *John Marota, Thomas Fairclough, Tom Antista* DESIGNER: *Thomas Fairclough* PHOTOGRAPHY: *Michael West Photography* AGENCY: *Antista Fairclough Design* CLIENT: *Anheuser Busch Inc.* COUNTRY: *USA* ■ The packaging was developed from 1800s resource material from the Anheuser Busch archives. This brand was developed to compete with microbreweries. ● *Diese Verpackung wurde auf der Basis von Archivmaterial der Brauerei aus der Zeit um 1800 entwickelt. Die Marke wurde lanciert, um den Produkten der kleinen Spezialbrauereien etwas entgegenzusetzen.* ▲ *Cet emballage a été créé sur la base des archives de la brasserie qui remontent aux années 1800. Cette marque a été lancée pour concurrencer les produits des petites brasseries.*

PAGE 21, *image 10* ART DIRECTOR: *John Marota (Anheuser Busch, Inc.), Thomas Fairclough, Tom Antista* DESIGNER: *Thomas Fairclough* PHOTOGRAPHER: *Michael West Photography* AGENCY: *Antista Fairclough* CLIENT: *Anheuser Busch* COUNTRY: *USA* ■ The packaging was developed from 1800s resource material from the Anheuser Busch archives. This brand was developed to compete with microbreweries. ● *Diese Verpackung wurde auf der Basis von Archivmaterial*

der Brauerei aus der Zeit um 1800 entwickelt. Die Marke wurde lanciert, um den Produkten der kleinen Spezialbrauereien etwas entgegenzusetzen.* ▲ *Cet emballage a été créé sur la base des archives de la brasserie qui remontent aux années 1800. Cette marque a été lancée pour concurrencer les produits des petites brasseries.*

PAGE 21, *image 11* ART DIRECTOR: *John Marota, Thomas Fairclough, Tom Antista* DESIGNER: *Thomas Fairclough* PHOTOGRAPHY: *Michael West Photography* AGENCY: *Antista Fairclough Design* CLIENT: *Anheuser Busch Inc.* COUNTRY: *USA* ■ This packaging design uses festive colors and shapes to convey the holiday spirit. ● *Mit Hilfe von festlich wirkenden Farben und Formen sollte mit dieser Verpackung für die Feiertage das Gefühl eines handgemachten Produktes erzeugt werden.* ▲ *Les couleurs de fête et les formes de cet emballage reflètent l'ambiance des jours fériés et des fêtes de fin d'année.*

PAGE 21, *image 12* ART DIRECTOR: *John Marota, Thomas Fairclough, Tom Antista* DESIGNER: *Thomas Fairclough* PHOTOGRAPHY: *Michael West Photography* AGENCY: *Antista Fairclough Design* CLIENT: *Anheuser Busch Inc.* COUNTRY: *USA* ■ Elephant Red was developed to compete with red lager beers. Power and high alcohol content were conveyed through the graphic presentation. ● *Elephant Red muss sich im Markt der dunklen Lagerbiere behaupten. Der kräftige Geschmack und der hohe Alkoholgehalt finden Ausdruck in der graphischen Gestaltung.* ▲ *Elephant Red a été conçue pour concurrencer les bières blondes-rousses. Sa conception graphique souligne son goût puissant et sa forte teneur en alcool.*

PAGE 21, *image 13* ART DIRECTOR: *Jack Anderson* DESIGNERS: *Jack Anderson, Larry Anderson, Bruce Branson-Meyer* PRODUCT PHOTOGRAPHER: *Tom McMackin* ARTIST/ILLUSTRATOR: *Mark Summers* AGENCY: *Hornall Anderson Design Works, Inc.* CLIENT: *William & Scott Company* COUNTRY: *USA* PRINTER: *6 pack: Zumbiel; Labels: Inland Printing; Caps: Zapata* TYPEFACE: *Copperplate, Reckleman, hand-lettering.* ■ The packaging design encompasses six-pack carrier cartons, bottles, and labels for Rhino Chasers beer. ● *Tragkartons, Flaschen und Etiketten für ein Bier der Marke Rhino Chasers.* ▲ *Boîtes en pack de six, bouteilles et étiquettes de la marque de bière Rhino Chasers.*

PAGE 21, *image 14* ART DIRECTOR/DESIGNER: *Amy Leppert* AGENCY: *Murrie Lienhart Rysner* CLIENT: *Kettle Moraine Beverage Co.* COUNTRY: *USA* ■ The goal was to create packaging with contemporary adult imagery for a new, clean malt beverage. ● *Hier ging es um die Schaffung eines zeitgemässen, ausgereiften Images für ein neues, reines Malzgetränk.* ▲ *Cet emballage avait pour but de créer une image contemporaine, «posée», pour une nouvelle bière pur malt.*

PAGE 21, *image 15* ART DIRECTORS: *Peter Di Donato, Don Childs* DESIGNER: *Don Childs* ARTIST/ILLUSTRATOR: *Rich Lo* AGENCY: *DiDonato Associates* CLIENT: *Goose Island Beer Company* COUNTRY: *USA* PRINTER: *Inland Label/Zumbiel Carton* TYPEFACE: *Trajan, Copperplate, Kunstlerscript, Modern No. 20* ■ Created for the retail launch of the product, this packaging is intended to communicate craft-brewed quality, to establish memorable shelf presence, and to create a system for product line expansion. ● *Lancierung eines neuen Produktes. Es ging um seine Positionierung als Qualitätsbier, um seine wirkungsvolle Präsentation im Verkaufsgestell und um die Schaffung eines graphischen Systems, das den Ausbau der Produktlinie erlaubt.* ▲ *Lancement d'une nouvelle bière sur le marché. Il s'agissait de la positionner en tant que bière de qualité supérieure, d'établir sa position dans les rayons et de créer un concept graphique qui permettrait d'élargir la gamme de produits.*

PAGE 21, *image 16* ART DIRECTOR: *Michael Osborne* DESIGNER: *Christopher Lehmann* PRODUCT PHOTOGRAPHER: *Tony Stromberg* ARTIST/ILLUSTRATOR: *Christopher Lehmann* AGENCY: *Michael Osborne Design* CLIENT: *St. Stan's Brewing Co.* COUNTRY: *USA* PRINTER: *Carrier: Crown Packaging* LABEL: *Louis Roesch Co. San Francisco* TYPEFACE: *Stymie, Copperplate* ■ Packaging created for the client's newest product and brand extension. ● *Lancierung einer neuen Biersorte als Erweiterung des Brauereisortiments.* ▲ *Emballage créé pour le dernier-né des produits du client, une extension de la marque.*

PAGE 22, *image 17* ART DIRECTOR/DESIGNER: *Fred Tieken* ILLUSTRATOR: *Tad A. Smith* PRODUCT PHOTOGRAPHER: *Paul Markow* AGENCY: *Tieken Design and Creative Services* CLIENT: *Black Mountain Brewing Company* COUNTRY: *USA* ■ The design is meant to suggest an afternoon in a Mexican cantina. The lime green background suggests the traditional lime wedge taken with Mexican beer. ● *Mexikanische Lebenslust, die Atmosphäre eines angenehmen Nachmittags in einer mexikanischen Cantina war das Thema dieser Packungsgestaltung. Der limonengrüne Hintergrund ist eine Anspielung auf das Stückchen Limone, das zum mexikanischen Bier serviert wird.* ▲ *Le graphisme de cet emballage devait transmettre une image gaie d'un après-midi passé dans une cantina mexicaine. Le fond vert jaune suggère le citron vert posé sur le bord du verre que l'on sert avec de la bière mexicaine.*

PAGE 22, *image 18* ART DIRECTOR/DESIGNER/ARTIST/ILLUSTRATOR: *Woody Pirtle* AGENCY: *Pentagram Design* CLIENT: *Flying Fish Brewing Co.* COUNTRY: *USA* TYPEFACE: *Matrix* ■ To establish a distinctive identity in a microbrewery market saturated with traditional imagery, this label was designed to be both modern and whimsical. ● *Um der kleinen Spezialbrauerei einen eigenen Auftritt zu ver-*

schaffen, sollte das Etikett modern und ganz speziell aussehen, im Gegensatz zu den auf Tradition getrimmten Etiketten, die in diesem Marktsegment so häufig verwendet werden. ▲ L'étiquette se devait d'être moderne et originale pour conférer une identité détonante à une petite brasserie et pour qu'elle se démarque de l'imagerie traditionnelle utilisée dans ce segment de marché saturé.

PAGE 22, *image 19* Art Directors: *John Marota, Thomas Fairclough, Tom Antista* Designer: *Thomas Fairclough Design Design* Photographer: *Michael West Photography* Agency: *Antista Fairclough* Client: *Anheuser Busch* Country: *USA* ■ *This packaging was developed from 1800s resource material from the Anheuser Busch archives. This brand was developed to compete with microbreweries.* ● *Die Verpackung wurde auf der Basis von Archivmaterial der Brauerei aus der Zeit um 1800 entwickelt. Diese Marke wurde lanciert, um den Produkten der kleinen Spezialbrauereien etwas entgegenzusetzen.* ▲ *Cet emballage a été créé sur la base des archives de la brasserie qui remontent aux années 1800. Cette marque a été lancée pour concurrencer les produits des petites brasseries.*

PAGE 22, *image 20* Art Directors: *John Marota, Thomas Fairclough, Tom Antista* Designer: *Thomas Fairclough Design* Photographer: *Michael West Photography* Agency: *Antista Fairclough* Client: *Anheuser Busch* Country: *USA* ■ *This packaging emphasizes "Bud Ice" as opposed to "Budweiser" Ice. New, contemporary textures and colors are utilized.* ● *Die Biersorte "Bud Ice" ist leichter als "Budweiser Ice", und darum geht es bei dieser Verpackung, die sich in Farbe und Gestaltung modern gibt.* ▲ *Cet emballage met l'accent sur "Bud Ice" au lieu de "Budweiser Ice". De nouvelles textures et couleurs contemporaines ont été utilisées.*

PAGE 22, *image 21* Art Director: *Glenn Tutssel* Designer: *Garrick Hamm* Agency: *Tutssels* Client: *Guinness Brewing UK* Country: *England* ■ *The Guinness Beer Font was designed to promote the draught product in all parts of the UK.* ● *Die "Guinness-Beer"-Schrift wurde entworfen, um für dieses Bier zu werben, das überall in Grossbritannien vom Hahn gezapft wird* ▲ *La typographie pour la bière Guinness a été créée pour la promotion de cette bière à pression dans toutes les régions du Royaume-Uni.*

PAGE 22, *image 22* Creative Director: *Michael Jager* Art Director: *David Covell* Designer: *David Covell* Agency: *Jager Di Paola Kemp Design* Client: *Magic Hat Brewing Company* Country: *USA* Printer: *General Press* Typeface: *Xavier, Minion, and Scala Sans* ■ *A striking graphic identity was created out of a moonface icon to evoke mystery and intrigue for the client's flagship beer.* ● *Dieses Bier ist das Flaggschiff des Kunden. Die spezielle Illustration soll dem Produkt etwas Geheimnisvolles verleihen.* ▲ *Une identité frappante a été créée à partir d'une lune pour conférer une part de mystère à cette bière, marque principale du client.*

PAGE 22, *image 23* Art Directors: *John Marota, Thomas Fairclough, Tom Antista* Designer: *Thomas Fairclough* Photography: *Michael West Photography* Agency: *Antista Fairclough Design* Client: *Anheuser Busch Inc.* Country: *USA* ■ *A mixture of old and new design elements were employed to convey the qualities of a rich, sophisticated beer.* ● *Eine Mischung von alten und neuen Gestaltungselementen vermittelt einen Eindruck der Qualität dieses vollen, anspruchsvollen Biers.* ▲ *Un mélange entre ancien et nouveau design a été utilisé pour mettre en valeur les qualités d'une bière riche et sophistiquée.*

PAGE 22, *image 24* Art Directors: *Jerry Anderlin (Hal Riney & Partners), Primo Angeli* Designer/Typographer: *Mark Jones* Artist/Illustrator: *Bruce Wolfe* Agency: *Primo Angeli* Client: *Hal Riney & Partners* Country: *USA* ■ *The objective was to create an identity for a regional microbrewery.* ● *Hier ging es um die visuelle Identität einer kleinen, regionalen Brauerei.* ▲ *Emballage ayant pour but de créer une identité visuelle pour une petite brasserie régionale.*

PAGE 22, *image 25* Art Directors: *John Marota, Thomas Fairclough, Tom Antista* Designer: *Thomas Fairclough Design* Photographer: *Michael West Photography* Agency: *Antista Fairclough* Client: *Anheuser Busch Inc.* Country: *USA* ■ *The label and packaging of this beer were developed with microbrew cues in order to position the product against beers of microbreweries.* ● *Bei der Verpackungsgestaltung ging es um die Abgrenzung des Biers gegen die Erzeugnisse von Mini-Brauereien.* ▲ *Cet emballage doit démarquer cette bière des produits des petites brasseries.*

PAGE 23, *image 26* Designer: *Christopher Hadden* Artist/Illustrator: *Marty Braun* Agency: *Christopher Hadden Design* Client: *Shipyard Brewing Co.* Country: *USA*

PAGE 23, *image 27* Art Director/Artist/Illustrator: *Rain Pikand* Designer: *Andrus Lember* Product Photographer: *Jaak Kadak* Agency: *Division* Client: *Saku Brewery, Ltd.* Country: *Finland* Printer: *Lauttasaaren Paino* Typeface: *Handlettering, M Grotesk, Engravers Gothic* ■ *This packaging was created for an Estonian brewery which produces specialty beers for events, festivals, or seasons. Saku Porter was produced for the Christmas season.* ● *Packungsgestaltung für eine estländische Brauerei, die Bier für spezielle Anlässe, Jahreszeiten und Feiertage herstellt. Saku Porter war für die Weihnachtszeit bestimmt.* ▲ *Emballage conçu pour une brasserie en Estonie qui produit des bières pour des occasions spéciales, des fêtes, des festivals ou selon les saisons. La bière Saku Porter a été créée pour les fêtes de fin d'année.*

PAGE 24, 25; *images 28, 31* Art Director: *Keizo Matsui* Designers: *Keizo Matsui, Yuko Araki* Agency: *Keizo Matsui & Associates* Client: *Yagi Shizou-bu* Country: *Japan* Printer: *Kotobuki Seiban Co., Ltd.* ■ *This bottle and label design was created for a local Japanese sake company.* ● *Flaschenausstattung für einen japanischen Hersteller von Sake.* ▲ *Habillage de bouteille pour un producteur japonais de saké.*

PAGE 25, *image 29* Art Director: *Akio Okumura* Designer: *Mitsuo Ueno* Agencies: *Packaging Create, Dentsu Inc. Kansai* Client: *Gekkeikan* Country: *Japan* Typeface: *Univers* ■ *This packaging was created for a Japanese sake.* ● *Packungsgestaltung für einen japanischen Sake.* ▲ *Emballage conçu pour un saké japonais.*

PAGE 25, *image 30* Art Director: *Akio Okumura* Designer: *Emi Kajihara* Agencies: *Packaging Create, Dentsu Inc.* Client: *Gekkeikan* Country: *Japan* Typeface: *Original* ■ *This packaging was created for Japanese sake* ● *Packungsgestaltung für einen japanischen Sake.* ▲ *Emballage conçu pour un saké japonais.*

PAGE 25, *image 32* Designer: *Toshio Kamitani* Artist/Illustrator: *Toshio Kamitani* Agency: *TCD Corporation* Client: *Hakutsuru Sake Brewing Co., Ltd.* Country: *Japan* Typeface: *Mincyo (MM-A-OKL)* ■ *The packaging targets young women and is designed to reflect a sense of harmony with Japan's seasons.* ● *Aufgabe dieser Packungsgestaltung ist es, junge Frauen anzusprechen und ein Gefühl harmonischen Einklangs mit den Jahreszeiten zu vermitteln.* ▲ *Cet emballage qui s'harmonise avec les saisons au Japon a pour public cible des jeunes femmes.*

PAGE 26, *image 33* Art Director: *Barrie Tucker* Designers/Artists/Illustrators: *Barrie Tucker, Jody Tucker* Product Photographer: *Simon Vaughn* Agency: *Tucker Design* Client: *Saddlers Creek Winery* Country: *Australia* Printer: *Label Leaders South Australia* Typeface: *Arcadia, Uncial Script* Paper: *Jac Wlk 202 - Self-Adhesive* ■ *This packaging was created for a premium red wine from Hunter Valley.* ● *Packungsgestaltung für einen erstklassigen australischen Wein aus dem Hunter Valley.* ▲ *Cet emballage a été conçu pour un vin rouge australien de la Hunter Valley.*

PAGE 26, *image 34* Art Director: *Barrie Tucker* Designers: *Barrie Tucker, Jody Tucker* Product Photographer: *Simon Vaughn* Artist/Illustrator: *Marie Smith* Agency: *Tucker Design* Client: *Negociants* Country: *Australia, New Zealand* Printer: *Hally Print New Zealand* Typeface: *Handscript, Helvetica* Paper: *Jac Wlk 202 - Self-Adhesive* ■ *The packaging was designed for a manufacturer and negociant of premium New Zealand wines.* ● *Die Verpackung wurde für einen Weinproduzenten und -händler von hochklassigen neuseeländischen Weinen entworfen.* ▲ *L'emballage a été conçu pour un producteur et négociant en vins néo-zélandais.*

PAGE 27, *images 35-38* Art Director: *Michel Logoz* Designer: *Jacques Zanoli* Photographer: *Magali Koenig* Illustrator: *Beat Brusch* Agency: *Creation Communication* Client: *Provins Valais* Country: *Switzerland* ■ *Bottle design for wines from the Valais, Switzerland.* ● *Flaschenausstattung für eine Reihe von Weinen aus dem Wallis, Schweiz.* ▲ *Habillage de bouteilles pour des vins du Valais.*

PAGE 28, *image 39* Art Director: *Michael Osborne* Designer: *Kristen Clark* Product Photographer: *Tony Stromberg* Agency: *Michael Osborne Design* Client: *Cornerstone Cellars* Printer: *Sheck Printing* Typeface: *Trajan* ■ *This wine label was created for the initial offering from a newly established winery.* ● *Etikett für das Einführungsangebot eines neu etablierten Weinguts.* ▲ *Cette étiquette a été créée pour les premiers produits proposés par un nouvel établissement vinicole.*

PAGE 28, *image 40* Art Director: *Paola Dashwood* Designer: *Derek Ventling* Product Photographer: *David Ogden* Artist/Illustrator: *Gary Sullivan* Agency: *Dashwood Design* Client: *Montana Wines* Country: *New Zealand* Printer: *Panprint, Zanzibar* Paper: *90 Chromolux plain* ■ *The packaging design was meant to be dynamic, contemporary, and bold in order to attract young wine drinkers.* ● *Das dynamische, moderne Design dieser Verpackung richtet sich vor allem an junge Weinliebhaber.* ▲ *Audace, dynamisme et design moderne, tels sont les critères qui ont prévalu à la conception de cet emballage. Groupe cible: les jeunes amateurs de vins.*

PAGE 29, *image 41* Designer: *Lynda Warner* Agency: *Lynda Warner, Graphic Design* Client: *Chateau Xanadu* Paper: *Matt Self-Adhesive* Typeface: *Futura Extra Bold* ■ *The design for this label takes inspiration from the Viennese modern art movement of the early 1900s.* ● *Vom Wiener Sezessionsstil inspiriertes Weinetikett.* ▲ *Etiquette de vin inspirée du style de la Wiener Sezession.*

PAGE 29, *image 42* Art Director: *John Blackburn* Designer: *Belinda Duggan* Artist/Illustrator: *Matt Thompson* Agency: *Blackburn's Limited* Client: *Cockburn Smithes & Cia Lda* Typeface: *Clarendon* ■ *This label was designed to help people understand Vintage Port, demystifying the product's exclusivity. The road sign icons were intended to communicate complex product information in a simple way* ● *Das Etikett klärt die Konsumenten über alten Portwein auf und entstaubt das Image dieses exklusiven Produktes. Mit Hilfe der Motive von Verkehrsschildern werden komplexe Informationen vermittelt.* ▲ *Cette étiquette éclaire les consommateurs sur le porto vieux (vintage) tout en démythifiant* *l'image de ce produit exclusif. Grâce aux motifs des panneaux de signalisation, des informations complexes sont transmises sur le produit.*

PAGE 29, *image 43* ART DIRECTOR: *Supon Phornirunlit* DESIGNER: *Andrew Berman* AGENCY: *Supon Design Group* CLIENT: *Grand Palace Food International* COUNTRY: *Thailand* PRINTER: *BKK Press* TYPEFACE: *Future* PAPER: *Crack-n-Peel*

PAGE 29, *image 44* AGENCY: *Cato Design* CLIENT: *T'Gallant* COUNTRY: *Australia* PRINTER: *Rothfield Print Management* TYPEFACE: *Revival* ■ This label design uses an unconventional die-cut in contrast with the traditional nature of this Pinot Grigio. ● *Ein unkonventionelles Etikett für den traditionsreichen Pinot Grigio.* ▲ *Une étiquette originale pour ce pinot gris riche en tradition*

PAGE 29, *image 45* AGENCY: *Cato Design* CLIENT: *T'Gallant* COUNTRY: *Australia* PRINTER: *Rothfield Print Management* TYPEFACE: *Revival* ■ This half-bottle was designed for a dessert wine produced in limited quantities. The label is intended to convey the wine's "richness." ● *Diese Halbliterflasche enthält einen Dessertwein, von dem nur kleine Mengen hergestellt werden. Das Etikett sollte die Vollmundigkeit des Weines reflektieren.* ▲ *Cette bouteille d'un demi-litre contient un vin de dessert produit en quantités limitées. L'étiquette devait évoquer la plénitude du vin.*

PAGE 29, *image 46* AGENCY: *Cato Design* CLIENT: *T'Gallant* COUNTRY: *Australia* PRINTER: *Rothfield Print Management* TYPEFACE: *Revival* ■ The double-sided label is designed to appeal to women. It features roses that are magnified when viewed through the bottle. ● *Dieses beidseitig bedruckte Etikett soll Frauen ansprechen. Durch die Flasche hindurch betrachtet, sieht man das Rosenmotiv wie durch ein Vergrösserungsglas.* ▲ *Cette étiquette imprimée recto verso s'adresse aux femmes. Si l'on regarde le motif de roses à travers la bouteille, on a l'impression de le voir à travers un verre grossissant.*

PAGE 29, *image 47* AGENCY: *Cato Design* CLIENT: *Mistwood Vineyard* COUNTRY: *Australia* PRINTER: *Rothfield Print Management* ■ The label design of this product changes yearly to reflect the attributes of that year's vintage. ● *Jeder Jahrgang dieses Weines erhält ein eigenes Etikett, das seine besonderen Eigenschaften reflektiert.* ▲ *Chaque année, l'étiquette de ce vin change afin de refléter les caractéristiques propres à chaque vendange.*

PAGE 29, *image 48* DESIGNER/ARTIST/ILLUSTRATOR: *Jeffrey Caldeway* AGENCY: *Caldeway Design* CLIENT: *Tom Eddy Napa Valley Wines* COUNTRY: *USA* PRINTER: *Gordon Graphics* TYPEFACE: *Lithos* PAPER: *Quest* ■ Predicated on the idea that great wines are products of the soil in which they are grown, pigments and texture of original art were created from crushed dirt and vine bark. ● *Aufbauend auf der Tatsache, dass grosse Weine ihre Güte der Erde verdanken, in der sie wachsen, wurden für den Originalentwurf des Etiketts natürliche Pigmente der Erde und die Rinde der Reben verwendet.* ▲ *L'idée-force de cet emballage est que les grands vins tirent leurs qualités du sol où ils ont été cultivés. Aussi, des pigments naturels du sol et l'écorce des ceps ont été utilisés pour le concept original.*

PAGE 29, *image 49* AGENCY: *Cato Design* CLIENT: *T'Gallant* COUNTRY: *Australia* PRINTER: *Rothfield Print Management* ■ In addition to being economical to produce, this label is meant to convey the product's strawberry flavor. ● *Thema des Etiketts, das sich kostengünstig herstellen lässt, ist der Erdbeergeschmack des Produktes.* ▲ *Le thème de cette étiquette, au prix de revient bon marché, rappelle le goût de fraises du produit.*

PAGE 30, *image 50* ART DIRECTOR: *Angelo Sganzerla* DESIGNER: *Angelo Sganzerla* ARTIST/ILLUSTRATOR: *Alfonso Goi* AGENCY: *Angelo Sganzerla* CLIENT: *Solci*

PAGE 30, *image 51* ART DIRECTOR/DESIGNER: *Patti Britton* PRODUCT PHOTOGRAPHER: *Mitch Rice* AGENCY: *Britton Design* CLIENT: *Viansa Winery* COUNTRY: *USA* PRINTER: *Bolling and Finke, Ltd.* TYPEFACE: *Futura Condensed* ■ A two-hundred year old obsidian arrowhead created by the American Indians is used to represent the client's celebration of California's past civilizations. ● *Die zweihundert Jahre alte Obsidian-Pfeilspitze der amerikanischen Indianer dient dem Winzer als Symbol der vergangenen Zivilisationen Kaliforniens für eine Spezialedition von Weinflaschen.* ▲ *Créée par les Indiens d'Amérique, cette pointe de flèche obsidienne sert de symbole pour représenter les anciennes civilisations de Californie.*

PAGE 30, *image 52* DESIGNERS: *Jeffrey Caldeway, Dan Maclain, Jim Murphy* ARTIST/ILLUSTRATOR: *Dan Maclain* AGENCY: *Caldeway Design* CLIENT: *Fetzer Vineyards* COUNTRY: *USA* PRINTER: *Bolling and Finke* TYPEFACE: *handlettering, copperplate* PAPER: *Trailblazer 100% kenaf tree-free vellum* ■ The torn, textured, hand lettered label and proprietary bottle shape reflect the natural origin of this organically produced product. ● *Etikett und Flaschenform reflektieren den biologischen Anbau dieses Weines.* ▲ *L'étiquette et la forme de la bouteille reflètent la culture biologique du vin.*

PAGE 30, *image 53* ART DIRECTOR: *Annette Harcus* DESIGNERS: *Kristen Thieme, Stephanie Martin, Annette Harcus* PRODUCT PHOTOGRAPHER: *Keith Arnold* ARTIST: *Paul Newton* AGENCY: *Harcus Design* CLIENT: *Yalumba Winery* COUNTRY: *Australia* PRINTER: *Collatype Labels* TYPEFACE: *Handdrawn, Future, Gill Sans* ■ The packaging design is meant to reflect the product's name, Antipodean, which means "places diametrically opposite to each other." Australia is often referred to

as "The Antipodes." ● *Der Name des Produktes (Antipode) war Thema der Packungsgestaltung für australischen Wein. Für viele Erdbewohner ist der Australier ein Antipode, ein Mensch, der auf der dem Betrachter gegenüberliegenden Seite der Erde wohnt.* ▲ *«Antipodean», le nom du produit, était le thème central pour la conception de cet emballage d'un vin australien: beaucoup de personnes estiment que l'Australie est aux antipodes de leur propre pays.*

PAGE 31, *image 54* DESIGNER/ARTIST/ILLUSTRATOR: *Jeffrey Caldeway* AGENCY: *Caldeway Design* CLIENT: *Goosecross Cellars* COUNTRY: *USA* PRINTER: *Gordon Graphics, V&V Metal Fabricators* TYPEFACE: *Garamond, Engravers Gothic* ■ The package design makes use of a metaphorical pun playing off the double entendre "Æros." Forged bronze and nickel wings form a sculptural framework for a cameo of the winged goddess Psyche, lover of Eros. ● *Bei dieser Packungsgestaltung geht es um den doppelten Sinn des Wortes Æros. Flügel aus Bronze und Nickel bilden den Rahmen für eine Miniatur der Göttin Psyche, der Geliebten von Eros.* ▲ *Cet emballage joue sur le double sens du mot «Æros». Les ailes en nickel et en bronze forment un cadre sculptural pour le camée de la déesse ailée Psyché, amante d'Eros.*

PAGE 32, *image 55* ART DIRECTOR/DESIGNER: *Avital Kellner Gazit* AGENCY: *Daedalos* CLIENT: *Golan Heights Winery* COUNTRY: *Israel*

PAGE 33, *image 56* ART DIRECTOR: *Barrie Tucker* DESIGNERS: *Barrie Tucker, Jody Tucker* DESIGNER: *Claire Rose* PRODUCT PHOTOGRAPHER: *Simon Vaughn* AGENCY: *Tucker Design* CLIENT: *Lactos* COUNTRIES: *Australia, New Zealand* PRINTER: *Five Star Press* TYPEFACE: *Stone Sans* PAPER: *CPA Mastertac Vellum Adhesive* ■ This silver package was developed for a Tasmanian cheese manufacturer's 40th anniversary. It features an image sandblasted into the bottle and a handcrafted, silver-plated stopper. ● *Diese silberne Verpackung wurde zum 40. Geburtstag eines Käseproduzenten aus Tasmanien entworfen. Das Bild wurde im Sandstrahlverfahren auf die Flasche übertragen, der versilberte Korken ist handgemacht.* ▲ *Cet emballage argenté a été créé pour le quarantième anniversaire d'un producteur de fromages en Tasmanie. L'image a été reportée sur la bouteille avec une sableuse; le bouchon argenté est fait main.*

PAGE 33, *image 57* ART DIRECTOR/DESIGNER: *Patti Britton* PRODUCT PHOTOGRAPHER: *M.J. Wickham* AGENCY: *Britton Design* CLIENT: *Viansa Winery* COUNTRY: *USA* PRINTER: *Bolling and Finke, Ltd.* TYPEFACE: *Augusta Inline, Centaur* ■ To accentuate the sculptural, elongated shape of the bottle, a segmented label was created—the upper label consists of a fresco in a classical renaissance-shaped inverted U form; the lower portion has a simple label for typography and the client's crest. ● *Um die skulpturale, schlanke Form der Flasche zu unterstreichen, wurden zwei Etiketts verwendet – das obere in Form eines umgedrehten U zeigt ein Fresko, das untere ist ein einfaches Etikett mit Schrift und dem Wappen des Herstellers.* ▲ *Pour accentuer la forme sculpturale et allongée de la bouteille, deux étiquettes ont été utilisées: celle du haut représente une fresque en forme d'un U retourné selon le style classique de la Renaissance; celle du bas est d'une grande simplicité dans sa typographie et montre l'emblème du client.*

PAGE 33, *image 58* ART DIRECTOR/DESIGNER: *Patti Britton* PRODUCT PHOTOGRAPHER: *Mitch Rice* AGENCY: *Britton Design* CLIENT: *Viansa Winery* COUNTRY: *USA* PRINTER: *Bolling and Finke, Ltd.* TYPEFACE: *Augusta Inline, Centaur* ■ The Italian tall flint bottle was chosen to complement the soft wine color. The label features a 15th century fresco from the walls of Viansa to re-create the Italian renaissance look. ● *Die schlanke italienische Feuerstein-Flasche wurde als ideale Ergänzung zur sanften Farbe des Weins gewählt. Auf dem Etikett sind Fresken des Ortes Viansa aus dem 15. Jahrhundert abgebildet, um die Aura der italienischen Renaissance heraufzubeschwören.* ▲ *Cette bouteille italienne en silex de forme allongée a été choisie parce qu'elle s'harmonise parfaitement avec la robe tendre du vin. L'étiquette montre une fresque du 15ème siècle peinte sur des murs de Viansa pour recréer l'atmosphère de la Renaissance italienne.*

PAGE 33, *image 59* ART DIRECTOR: *Barrie Tucker* DESIGNERS: *Barrie Tucker, Hans Kobla* PRODUCT PHOTOGRAPHER: *Simon Vaughn* AGENCY: *Tucker Design* CLIENT: *Woods Bagot* COUNTRY: *Australia* TYPEFACE: *Avenir* ■ This bottle was created as a sculptural form to celebrate the 125th anniversary of Woods Bagot Architects. ● *Diese Flasche wurde als Skulptur konzipiert, um das 125jährige Bestehen einer Architekturfirma zu feiern.* ▲ *Cette bouteille, conçue comme une sculpture, commémore le 125ème anniversaire de la société Woods Bagot Architects.*

PAGE 33, *image 60* ART DIRECTOR: *Barrie Tucker* DESIGNERS: *Barrie Tucker, Hans Kobla* ARTIST/ILLUSTRATOR: *Hans Kobla* PRODUCT PHOTOGRAPHER: *Simon Vaughn* AGENCY: *Tucker Design* CLIENT: *Southcorp Wines* COUNTRY: *Australia* PRINTER: *ASAP Adelaide* TYPEFACE: *Bodoni* ■ Packaging created for Seppelt brand Viva 1 & 2 Liqueur Shiraz and Liqueur Chardonnay . The design is printed onto the bottles. ● *Flaschengestaltung für zwei Liköre. Die Flaschen wurden direkt bedruckt.* ▲ *Emballage conçu pour deux liqueurs. Impression directe sur les bouteilles.*

PAGE 33, *image 61* ART DIRECTOR/ARTIST/ILLUSTRATOR: *Barrie Tucker* DESIGNERS: *Claire Rose, Barrie Tucker* PRODUCT PHOTOGRAPHER: *Simon Vaughn* AGENCY: *Tucker Design* CLIENT: *Lactos* COUNTRY: *Australia* JEWELLER/METALSMITH: *Peter Coombs* ■ This is a limited-edition bottle created as a Christmas gift. The three-dimensional star is gold-coated sterling silver with a moonstone at the center of *the eye-graphic, the design firm's symbol. ● In limitierter Anzahl hergestellte Flasche, die als Weihnachtsgeschenk gedacht war. Der dreidimensionale Stern besteht aus vergoldetem Sterling-Silber mit einem Mondstein im Zentrum des Auges, dem Symbol der Designfirma. ▲ Bouteille en série limitée créée comme cadeau de Noël. L'étoile tridimensionnelle est en argent fin doré avec, au centre de l'œil, une pierre de lune, symbole de l'agence de design.*

PAGE 34, *image 62* ART DIRECTORS: *Steve Mitchell, Bill Thorburn* DESIGNER: *Chad Hagen* COPY: *Matt Elhardt* AGENCY: *Thorburn Design* CLIENT: *Millenium* COUNTRY: *Germany* TYPEFACE: *Trade Gothic* ■ *The packaging contains a liquor product with the identity of an industrial oil. ● Ein Kräuterlikör, verpackt wie ein industrielles Öl. ▲ Liqueur aux herbes avec un emballage rappelant celui d'une huile pour voiture.*

PAGE 35, *image 63* ART DIRECTOR: *Felix Harnickell* DESIGNER: *Claudia Wilken* PRODUCT PHOTOGRAPHER: *Stephan Försterling* ILLUSTRATOR: *Dietrich Ebert* AGENCY: *Baxmann & Harnickell* CLIENT: *H.C. Asmussen* COUNTRY: *Germany* PRINTER: *Offset Glücksburg* TYPEFACE: *Copperplate, Bembo, English script type* ■ *New packaging design for a liquor. ● Neugestaltung der Verpackung für einen Aquavit. ▲ Nouvel emballage pour de l'aquavit.*

PAGE 35, *image 64* ART DIRECTOR: *Barrie Tucker* DESIGNERS: *Barrie Tucker, Hans Kohla, Nick Mount* PRODUCT PHOTOGRAPHER: *Simon Vaughn* AGENCY: *Tucker Design* CLIENT: *Spicers* COUNTRY: *Australia* BOTTLE MANUFACTURE: *JAM Factory Adelaide* PRINTER: *Gift box: Stalley Box Co; Box Sticker: Five Star Press* TYPEFACE: *Gill Sans* PAPER: *Gift box: Mirrorkote; Box sticker: Spicers self-adhesive* ■ *This item was created for a paper merchant as a corporate Christmas gift. Hand-blown glass bottles were filled with wood-aged fortified Chardonnay. ● Handgefertigte Glasflaschen mit Verpackung als Weihnachtsgeschenk für die Kunden eines Papierherstellers. Der Inhalt: in Holzfässern gereifter Chardonnay. ▲ Bouteilles en verre soufflé contenant du chardonnay vieilli en fûts avec emballage comme cadeau de Noël pour les clients d'un fabricant de papier.*

PAGE 36, *image 65* ART DIRECTOR/DESIGNER: *Taku Satoh* PRODUCER: *Unison Network* CLIENT: *Takara Shuzo Co., Ltd.* COUNTRY: *Japan* ■ *This simple bottle design was created to reflect the modern distillation process used in making the product. ● Die Schlichtheit dieser Flasche reflektiert das moderne Destillationsverfahren bei der Herstellung dieses Produktes. ▲ La simplicité de cette bouteille reflète le processus de distillation moderne utilisé pour la fabrication de ce produit.*

PAGE 36, *image 66* ART DIRECTOR/DESIGNER: *Rob Verbaart* CLIENT: *Hooghoudt BV* COUNTRY: *Netherlands* PRINTER: *Söllner, Germany* TYPEFACE: *Customized Futura Display, Helvetica Extra Bold* ■ *The design of this bottle was inspired by the brand name and the associations linked to the drinking of Vodka. The blue and silver refer to its royal origins and to the crystalline clarity of ice. Affixed to the middle of the of the bottle is a coin featuring the portrait of the Dutch queen. ● Der Markenname und die mit Wodka verbundenen Assoziationen bestimmten die Gestaltung dieser Flasche. Blau- und Silber spielen sowohl auf seine königliche Herkunft als auf die kristalline Klarheit von Eis an. In der Mitte der Flasche ist eine Münze mit dem Porträt der niederländischen Königin angebracht. ▲ La marque de fabrique et les associations relatives à la vodka ont inspiré la conception de cette bouteille. Les tons bleus et argentés rappellent ses origines royales et la clarté cristalline de la glace. Une pièce de monnaie à l'effigie de la reine des Pays-Bas est fixée sur le milieu de la bouteille*

PAGE 37, *image 67* ART DIRECTOR/DESIGNER: *Gérard Billy* PRODUCT PHOTOGRAPHER: *Jacques Villegier* AGENCY: *Daedalus Design* CLIENT: *Societe Slaur* COUNTRY: *France* PRINTER: *Imprimerie du Bois de la Dame* TYPEFACE: *Hand-drawn, American* PAPER: *Centaure Ivoire Teinté* ■ *Complete bottle design (including labels, medaillion and stopper) for a 15-year-old whisky, targeted mainly at the French market. ● Vollständige Flaschenausstattung für einen 15 Jahre alten Whisky der gehobenen Preisklasse, der hauptsächlich für den französischen Markt bestimmt ist. ▲ Habillage complet d'une bouteille pour un whisky de choix de 15 ans d'âge (marché français essentiellement).*

PAGE 37, *image 68* ART DIRECTOR: *John Blackburn* DESIGNER: *Belinda Duggan* AGENCY: *Blackburn's LTD* CLIENT: *Berry Bros and Rudd Ltd.* COUNTRY: *UK* TYPEFACE: *Hand-drawn* ■ *The design objective was to create a radical and individualistic packaging identity in order to revive the brand. The bottle shape and hand-signed label are intended to communicate the authenticity of the malt. ● Aufgabe war die Schaffung einer aussergewöhnlichen, charakteristischen Verpackung, die dem Produkt zu neuem Leben verhelfen sollte. Die Flaschenform und das handsignierte Etikett betonen die Echtheit des Malt Whisky. ▲ Il s'agissait de créer un emballage inhabituel et marquant afin de donner un nouveau souffle au produit. La forme de la bouteille et l'étiquette signée à la main soulignent l'authenticité de ce whisky au malt.*

PAGE 37, *image 69* ART DIRECTOR/DESIGNER: *Sibylle Haase* AGENCY: *Atelier Haase & Knels* CLIENT: *B. Grashoff Nachf.* COUNTRY: *Germany* PRINTER: *H.M. Hauschild GmbH* ■ *Design of the bottle label for an alcoholic beverage. ● Gestaltung eines Flaschenetiketts für ein alkoholisches Getränk ▲ Conception d'une étiquette de bouteille pour une boisson alcoolisée.*

PAGE 37, *image 70* ART DIRECTOR/DESIGNER: *Roger Akroyd* AGENCY: *Michael Peters Limited* CLIENT: *Courvoisier S.A. France* COUNTRY: *France* ■ *Understated graphics*

enhance the elegance of this unique bottle structure. The product is a new brand designed to compete against other cognac products in the Chinese and Taiwanese markets. ● Zurückhaltende Graphik unterstreicht hier die Eleganz der Flasche. Es geht um eine neue Cognac-Marke, die es mit der Konkurrenz auf dem chinesischen und dem taiwanesischen Markt aufnehmen muss. ▲ Le graphisme discret souligne l'élégance de la bouteille. Il s'agit d'une nouvelle marque de cognac lancée sur les marchés chinois et taïwanais.

PAGE 38, *image 71* DESIGNER: *Alan Colvin* AGENCY: *Duffy Design* CLIENT: *Jim Beam* COUNTRY: *USA* ■ *This project was commissioned to celebrate the 200th anniversary of Jim Beam. The decanter was designed for a special bourbon and is contained in a gift box that uses elements from the product's history and the decanter design. ● Zum 200jährigen Bestehen von Jim Beam entworfene Geschenkbox für einen besonderen Bourbon. Für die Box wurden Elemente von Jim Beams Geschichte und der Flaschengestaltung verwendet. ▲ Emballage-cadeau contenant un bourbon spécial conçu à l'occasion du 200ème anniversaire de Jim Bean. L'emballage présente des éléments de l'histoire du produit et de son design.*

PAGE 38, *image 72* ART DIRECTOR: *John Blackburn* DESIGNER: *Belinda Duggan* ARTIST/ILLUSTRATOR: *Matt Thompson* AGENCY: *Blackburn's Limited* CLIENT: *United Distillers* COUNTRY: *Mexico* TYPEFACE: *Folio Bold Condensed* ■ *The packaging was created to emphasize the rum's Amazonian origins. The holographic chameleon changes color and was designed for an upscale market. The chameleon is also embossed scampering up the back of the bottle. ● Die Herkunft des Rums aus dem Amazonasgebiet lieferte das Thema für diese Verpackung. Das holographische Chameleon verändert seine Farbe und wird auf der Rückseite als Gravur im Glas wiederholt. Das Produkt gehört zur gehobenen Preisklasse. ▲ Les origines amazoniennes du rhum sont le thème central de cet emballage. Le caméléon holographique change de couleur et est reproduit sous forme de gravure sur le dos de la bouteille. Produit haut de gamme.*

PAGE 39, *image 73* ART DIRECTOR/DESIGNER: *Gérard Billy* ILLUSTRATOR: *Cyrille Bartolini* PRODUCT PHOTOGRAPHER: *Jacques Villegier* AGENCY: *Daedalus Design* CLIENT: *Cognac Raymound Ragnaud* COUNTRY: *France* PRINTER: *For the carton, label and collar: Imprimerie Bru Jarnac, France* TYPEFACE: *Hand-drawn, Caslon* PAPER: *Polyester Argent (carton), Centaure Ivoire Teinté (labels)* ■ *Complete packaging concept for a cognac targeted primarily at the Asian market. ● Vollständiges Verpackungskonzept für einen erstklassigen Cognac (Grande Champagne), der vor allem für den asiatischen Markt bestimmt ist. ▲ Concept de packaging complet pour un cognac de premier choix (Grande Champagne), destiné essentiellement au marché asiatique.*

PAGE 39, *image 74* ART DIRECTOR: *Georges Lachaise* DESIGNER: *Henri Champy* PRODUCT PHOTOGRAPHERS: *Studio Appi, Klaus Ramshorn* AGENCY: *George Lachaise Design* CLIENT: *Elie-Arnaud Denoix* COUNTRY: *France* PRINTER: *Georges Lachaise* ■ *Package design for the launch of a new brand of liquor. ● Packungsgestaltung für die Einführung eines Schnapses. ▲ Lancement pour une nouvelle marque d'eau-de-vie.*

PAGE 40, *image 75* ART DIRECTOR/DESIGNER: *Lucie Billy* ILLUSTRATOR: *Cyrille Bartolini* PRODUCT PHOTOGRAPHER: *Jacques Villegier* AGENCY: *Daedalus Design* CLIENT: *France Euro Agro* COUNTRY: *France* PRINTERS: *for the carton Groupe Otor, imprimerie Etuis Cognac; for the screen on the bottle Société Sobodoc* TYPEFACE: *Algérian retravaillé* PAPER: *support pour l'étui Polyester Aluminise Argent* ■ *Complete packaging including bottle design for the launch of a premium vodka in the Polish market. ● Vollständige Packungsgestaltung einschliesslich der Flasche für die Lancierung eines erstklassigen Wodkas auf dem polnischen Markt. ▲ Ici, il s'agissait de concevoir l'emballage ainsi que la bouteille pour le lancement d'une vodka de choix sur le marché polonais.*

PAGE 40, *image 76* ART DIRECTOR/DESIGNER: *Gérard Billy* ILLUSTRATOR: *Cyrille Bartolini* PRODUCT PHOTOGRAPHER: *Jacques Villegier* AGENCY: *Daedalus Design* CLIENT: *Cognac Raymound Ragnaud* COUNTRY: *France* PRINTER: *(for carton) Imprimerie Bru (pour la sérigraphe) Société Sobodoc* TYPEFACE: *Hand-drawn, Caslon* PAPER: *Support for the Carton: Polyester Aluminise Argent, Bottle neck collar: Centaure Ivoire Teinté* ■ *Complete package design for a cognac targeted at the Asian market. ● Vollständige Packungsgestaltung für einen Cognac (1er Cru, Grande Champagne), der hauptsächlich für den asiatischen Markt bestimmt ist. ▲ Emballage conçu pour un cognac (1er Cru, Grande Champagne), principalement destiné au marché asiatique.*

PAGE 41, *image 77* ART DIRECTORS: *Kristin Sommese, Lanny Sommese* DESIGNER: *Kristin Sommese* PHOTOGRAPHY: *Penn State Photographic Service* ARTIST/ILLUSTRATOR: *Lanny Sommese* AGENCY: *Sommese Design* CLIENT: *Aquapenn Spring Water Co.* COUNTRY: *USA* PRINTER: *Hammer Litho* TYPEFACE: *Matrix* ■ *This goal of this design was to convey elements of perfection, style, purity, and excellence as symbolized by the black panther. ● Hier ging es um die Darstellung von Eigenschaften wie Perfektion, Stil, Reinheit und Klasse, für die der schwarze Panther als Symbol dient. ▲ Le design devait faire transparaître des qualités, telles que perfection, style, pureté et classe, symboles de la panthère noire.*

PAGE 41, *image 78* ART DIRECTOR/DESIGNER: *Antonella Trevisan* ILLUSTRATOR: *Sergio Quaranta* AGENCY: *Tangram Strategic Design* CLIENT: *Gruppo Sun-Assago* COUNTRY:

Italy PRINTER: *Litografia Seven Verona, Italy* TYPEFACE: *Gil Sans Monotype, Gil Sans Condensed* ▪ The purpose was to improve the product's image through a clear and strong range identity, including suggested use at parties. ● *Mit einem klaren, starken Auftritt sollte das Produkt gefördert und seine Eignung als Party-Getränk unterstrichen werden.* ▲ *L'objectif de cet emballage était d'améliorer l'image du produit grâce à une forte identité, clairement définie. Il s'agissait également de suggérer que cette boisson peut être consommée lors de soirées.*

PAGE 42, *image 79* ART DIRECTOR: *Jack Anderson* DESIGNERS: *Jack Anderson, Julia Lapine, Heidi Favour, Leo Raymundo, Jill Bustamante* PRODUCT PHOTOGRAPHER: *Darrell Peterson* AGENCY: *Hornall Anderson Design Works, Inc.* CLIENT: *Talking Rain* COUNTRY: *USA* PRINTER: *Universal Sprayware* TYPEFACE: *Michelangelo* ▪ These bottles and cans were designed for the client's sparkling, fruit-flavored water. ● *Flaschen und Dosen für Fruchtgetränke.* ▲ *Conception de bouteilles et de boîtes pour une eau gazeuse fruitée.*

PAGE 42, *image 80* ART DIRECTORS: *Carla Pagoda, Brody Hartman, Primo Angeli* DESIGNERS: *Mark Jones, Brody Hartman* TYPOGRAPHER: *Sherry Bringham* ILLUSTRATOR: *Rick Gonella* AGENCY: *Primo Angeli* CLIENT: *G. Heileman Brewing Co.* COUNTRY: *USA* ▪ This packaging was designed to characterize the product as an old-fashioned root beer. ● *Bei dieser Verpackung ging es darum, das Produkt, eine Art Limonade, als das darzustellen, was es ist: das gute, alte amerikanische "Root Beer".* ▲ *Cet emballage devait conférer un caractère traditionnel à une sorte de limonade à base d'extraits végétaux dite «root beer».*

PAGE 43, *image 81* DESIGNER: *Alan Colvin* AGENCY: *Duffy Design* CLIENT: *The Coca-Cola Company* COUNTRY: *USA* PRINTER: *Mead* ▪ The curves and lines which are associated with Coca-Cola are utilized to further leverage brand equity. ● *Formen und Linien der typischen Coca-Cola-Flasche wurden eingesetzt, um den Markenauftritt zu unterstreichen.* ▲ *Les courbes et les lignes associées à Coca-Cola ont été utilisées pour mettre en valeur la marque.*

PAGE 44, *image 82* ART DIRECTORS: *Ellen Baker, Harumi H. Kubo* DESIGNER: *Harumi H. Kubo* PRODUCT PHOTOGRAPHER: *Ken Kubo* ARTIST/ILLUSTRATOR: *Robert Evans* CLIENT: *Nestle Beverage Co.* COUNTRY: *USA* PRINTER: *Reynolds Aluminum* TYPEFACE: *Lithos Family* ▪ This package is designed to expand the client's market reach. ● *Mit dieser Verpackung für ein Fruchtgetränk möchte der Kunde neue Märkte erschliessen.* ▲ *L'objectif de cet emballage de soda était de permettre au client d'élargir ses canaux de distribution.*

PAGE 45, *image 83* ART DIRECTOR: *Joe Howe, FCSD* DESIGNER: *Sue Hallworth* AGENCY: *Howe Design* CLIENT: *A.G. Barr p.l.c.* COUNTRY: *England* PRINTER: *Nacano Ltd* TYPEFACE: *Mac created, hand drawn* ▪ This packaging for cream soda was designed to give this product brand identity. ● *Bei dieser Verpackung für ein Erfrischungsgetränk ging es um die Etablierung der Markenidentität.* ▲ *Cet emballage pour une boisson rafraîchissante a été conçu pour établir l'identité de marque.*

PAGE 45, *image 84* ART DIRECTOR: *Kobe* DESIGNERS: *Kobe, Neil Powell, Alan Levsink* ARTISTS/ILLUSTRATORS: *Kobe, Neil Powell, Alan Levsink* AGENCY: *Duffy Design* CLIENT: *The Coca-Cola Company* COUNTRY: *USA* ▪ This soft drink packaging was created to appeal to youth worldwide. ● *Verpackung für ein Erfrischungsgetränk, dessen Zielgruppe Jugendliche in aller Welt sind.* ▲ *Emballage de soda qui s'adresse aux jeunes du monde entier.*

PAGE 45, *image 85* ART DIRECTOR: *Joe Duffy* DESIGNER/ARTIST/ILLUSTRATOR: *Jeff Johnson* AGENCY: *Duffy Design* CLIENT: *The Coca-Cola Company* COUNTRY: *USA* PRINTER: *Reynolds* ▪ After the product's baseline bottle design was created, it was applied to a variety of point-of-sale pieces and vending fronts. Since glass is a dangerous and impractical vending material, the design was altered and applied to the can. ● *Aus praktischen und Sicherheitsgründen entschied sich der Hersteller zugunsten von Dosen anstelle der ursprünglich vorgesehenen Flaschen. Die Gestaltung erstreckte sich auch auf verschiedene Laden-Displays.* ▲ *Pour des raisons pratiques et de sécurité, le fabricant a finalement préféré utiliser des boîtes au lieu des bouteilles prévues à l'origine. Le concept a également été appliqué à des présentoirs de magasin.*

PAGE 45, *image 86* ART DIRECTOR: *Joe Duffy* DESIGNER/ARTIST/ILLUSTRATOR: *Jeff Johnson* AGENCY: *Duffy Design* CLIENT: *The Coca-Cola Company* COUNTRY: *USA* PRINTER: *Reynolds* ▪ A new name and a variation of the mainline product design were employed to convey the idea of a, less-sweet, lower-caloried drink. ● *Um die Vorstellung eines frischeren und weniger süssen Produktes mit weniger Kalorien zu unterstützen, wurde ein neuer Name eingesetzt und das Design der etablierten Getränkelinie des Kunden abgewandelt.* ▲ *Nom et concept d'origine ont été revisités pour souligner les qualités d'une boisson plus légère, moins sucrée et moins riche en calories.*

PAGE 46, *images 87–89* ART DIRECTOR: *Tom Antista, Thomas Fairclough* PRODUCT PHOTOGRAPHY: *Michael West Photography* AGENCY: *Antista Fairclough Design* CLIENT: *Royal Crown Cola* COUNTRY: *USA* ▪ These promotional cans were developed for the client's core brand, highlighting the heritage of the client's name by utilizing an upscale design with a contemporary feel. ● *Für Promotionszwecke verschiedener Cola-Varianten entworfene Dosen, wobei der berühmte Markenname durch modernes, anspruchsvolles Design hervorgehoben wird.* ▲

Diverses boîtes conçues pour promouvoir la marque principale du client mise en valeur par un design moderne et haut de gamme.

PAGE 46, *image 90* ART DIRECTOR: *Paola Dashwood* DESIGNERS: *Paola Dashwood, Sarah Hingsion* PHOTOGRAPHER: *François Maritz* PRODUCT PHOTOGRAPHER: *David Ogden* ARTIST/ILLUSTRATOR: *Graham Downs* AGENCY: *Dashwood Design* CLIENT: *Frucor Beverages* COUNTRY: *New Zealand* PRINTER: *NZ Labels* TYPEFACE: *Hand-drawn type* PAPER: *Adhesive-Fasson Echo Gloss* ▪ The client is a major manufacturer of mineral water and juices in New Zealand. The packaging is designed to appeal to women 25 years and older. ● *Auftraggeber ist einer der grössten Produzenten von Mineralwasser und Säften in Neuseeland. Die Packung soll Frauen im Alter von über 25 Jahren ansprechen.* ▲ *Le client est un des principaux fabricants d'eau minérale et de jus de fruits en Nouvelle-Zélande. L'emballage a été conçu pour un groupe cible de femmes âgées de 25 ans ou plus.*

PAGE 46, *image 91* ART DIRECTORS: *Rex Peteet, Bryan Jessee* DESIGNER: *Derek Welch* PRODUCT PHOTOGRAPHER: *David Grimes* AGENCY: *GSD+M* CLIENT: *RC Cola/Royal Crown Company, Inc.* COUNTRY: *USA* TYPEFACES: *various & calligraphy* ▪ The packaging was designed to introduce a coffee soda beverage developed in response to the increasing demand and popularity of coffee and related refreshments. ● *Verpackung für ein Erfrischungsgetränk, das angesichts der steigenden Beliebtheit von Erfrischungen auf Kaffeebasis lanciert wurde.* ▲ *Emballage réalisé pour lancer une nouvelle boisson rafraîchissante à base de café en réponse à une demande croissante pour ce genre de produits.*

PAGE 46, *image 92* ART DIRECTOR: *Ralph J. Miolla* DESIGNER: *Robert J. Swan* ILLUSTRATOR: *Brian Sheridan* AGENCY: *Port Miolla Associates* CLIENT: *Great Brands of Europe* ▪ This package was designed for flavored premium spring water. ● *Packungsgestaltung für ein Qualitäts-Mineralwasser mit Geschmack.* ▲ *Emballage créé pour une eau minérale aromatisée de qualité supérieure.*

PAGE 46, *image 93* ART DIRECTOR: *Clemens Metzler* DESIGNER: *Clemens Metzler* PRODUCT PHOTOGRAPHER: *Volker Weinbold* AGENCY: *Fritzsch & Mackat, Werbeagentur* CLIENT: *Spreequell Mineralbrunnen GmbH* COUNTRY: *Germany* ▪ New design for an extended line of mineral water for the mass market. ● *Neugestaltung einer erweiterten Mineralwasser-Linie der unteren Preisklasse für grosse Absatzmärkte.* ▲ *Nouveau concept pour une vaste gamme d'eau minérale de grande distribution.*

PAGE 46, *image 94* ART DIRECTOR: *Joe Howe, FCSD* DESIGNER: *Chrissie Taylor* AGENCY: *Howe Design* CLIENT: *A.G. Barr p.l.c.* COUNTRIES: *England, Scotland* PRINTER: *Nacano, Ltd., Spowarts, Ltd.* TYPEFACE: *Mac created, hand-drawn* PAPER: *bottle labels on metallic paper* ▪ Red Kola has been a traditional flavor of Barr soft drinks for many years. It was decided to develop Red Kola into a brand in its own right. ● *Red Kola war viele Jahre lang der Inbegriff für die Geschmacksrichtung aller Erfrischungsgetränke des Herstellers. Mit dem neuen Auftritt sollte ein eigenständiges Image für das Getränk geschaffen werden.* ▲ *Depuis de nombreuses années, «red kola» est un arôme traditionnel utilisé par Barr pour ses boissons rafraîchissantes. La société a décidé de créer sa propre marque Red Kola.*

PAGE 46, *image 95* ART DIRECTOR: *Joe Howe, FCSD* DESIGNER: *Sue Hallworth* AGENCY: *Howe Design* CLIENT: *A.G. Barr p.l.c.* COUNTRY: *England* PRINTER: *Continental Can Europe* TYPEFACE: *Mac created, hand-drawn* ▪ Irn-Bru is a brand development in the sports drink market and is the company's leading brand. ● *Irn-Bru ist die Marke eines Getränkes für Sportler und auch das Flaggschiff des Herstellers.* ▲ *Irn-Bru, marque principale du fabricant, est une boisson destinée aux sportifs.*

PAGE 47, *image 96* ART DIRECTOR: *Jody Tucker* DESIGNERS: *Claire Rose, Jody Tucker* PRODUCT PHOTOGRAPHER: *Simon Vaughn* ILLUSTRATOR: *Keith McEwan* AGENCY: *Tucker Design* CLIENT: *Berrivale Orchards* COUNTRY: *Australia* PRINTER: *Labelmakers* TYPEFACE: *Caslon Open Face* PAPER: *Fasson Mirage* ▪ This package design was created for a range of carbonated spring waters with natural fruit flavors. ● *Packungsgestaltung für eine Reihe von Erfrischungsgetränken mit natürlichem Fruchtaroma.* ▲ *Emballage conçu pour des boissons rafraîchissantes aux arômes naturels.*

PAGE 48, *image 97* ART DIRECTOR: *Sophie Farbi* DESIGNERS: *Elia Hasbani, Franck Weidel* PRODUCT PHOTOGRAPHER: *Dominique Issermann* AGENCY: *Desgrippes Gobé & Associates* CLIENT: *Cartier* COUNTRY: *France* ▪ This bottle was designed to be simple, feminine, and modern. Evoking the client's background in jewelry, the bottle is encased in a gold-trimmed, red "jewel box." ● *Einfach, feminin und modern sollte dieser Flakon sein. Er ist in eine goldumrandete, rote «Schmuckschatulle» verpackt – eine Anspielung auf Cartiers Tradition als Juwelier.* ▲ *Ce flacon devait être simple, féminin et moderne. Il est emballé dans un «écrin» rouge avec une bordure dorée, une allusion à la vocation première de Cartier, la bijouterie/joaillerie.*

PAGE 49, 50; *images 98, 99* ART DIRECTORS: *Sally-Jane Costen, Lutz Herrmann* DESIGNER: *Veit Mahlmann* PHOTOGRAPHER: *Carl Lyttle* PRODUCT PHOTOGRAPHER: *Andreas Klingberg* AGENCY: *Peter Schmidt Studios* CLIENT: *Eurocos* COUNTRY: *Germany* ▪ This packaging was created to appeal to the same target consumer group as the fashion line associated with this fragrance. ● Mit dieser Verpackung für ein Herrenparfum soll die Zielgruppe angesprochen werden, an die sich auch die Hugo-Boss-Mode wendet. ▲ Cet emballage pour une eau de toilette pour hommes s'adresse au même groupe cible que les créations Hugo Boss.

PAGE 51, *image 100* ART DIRECTORS: *Thomas Fairclough, Tom Antista* DESIGNER: *Thomas Fairclough* ARTIST/ILLUSTRATOR: *Flatland* PHOTOGRAPHY: *Jody Dole, Michael West Photography* AGENCY: *Antista Fairclough Design* CLIENT: *Montsource, Inc.* COUNTRY: *USA* ■ The design utilizes a combination of classical sculpture, contemporary shapes and textures to create a strong masculine identity. ● Eine Kombination klassischer Skulptur und moderner Formen und Texturen sollte für ein starkes, männliches Image sorgen. ▲ Le design allie sculpture classique, formes et textures modernes pour créer une forte identité masculine.

PAGE 51, *image 101* DESIGNER: *Haley Johnson* PRODUCT PHOTOGRAPHER: *Paul Irmiter* AGENCY: *Haley Johnson Design Co.* CLIENT: *The Makeup Group* COUNTRY: *USA* TYPEFACE: *Trixie, Helvetica* PAPER: *Kraft*

PAGE 51, *image 102* ART DIRECTOR: *Hans D. Flink* DESIGNERS: *Chang–Mei Lin, Suzanne Clark* AGENCY: *Hans D. Flink Design Inc.* CLIENT: *Fabergé Co.* COUNTRY: *USA* ■ This packaging design uses an electric blue and black brand image to promote a new line of men's toiletries. ● Packungsgestaltung mit dem typischen leuchtenden Blau und Schwarz der Marke für die Einführung einer neuen Körperpflegereihe für Männer. ▲ Emballage bleu électrique et noire pour le lancement d'une nouvelle ligne de soins corporels pour hommes.

PAGE 51, *image 103* ART DIRECTOR: *Mary Scott* DESIGNERS: *Paul Farris, Winnie Li* PHOTOGRAPHER: *Toshio Nakajima* AGENCY: *Maddocks and Co* CLIENT: *Gryphon Development/Art deGaetano* COUNTRY: *USA* PRINTER: *Crickett Converters* TYPEFACE: *Trixie, Gill Sans, Gill Sans Bold* PAPER: *Vinyl (PVC labels)* ■ Pacifica Produce is a retail beauty and health care line that targets women aged 15-25. ● Pacifica Produce ist eine Kosmetiklinie, deren Zielgruppe junge Frauen zwischen 15 und 25 Jahren sind. ▲ La ligne de cosmétiques Pacifica Produce s'adresse à des jeunes femmes âgées de 15 à 25 ans.

PAGE 52, *image 104* ART DIRECTOR: *Doo H. Kim* DESIGNERS: *Dongil Lee, Seunghee Lee* AGENCY: *DooKim Design* CLIENT: *Utoo Zone, Samsung Corporation* COUNTRY: *Korea* ■ This design is based on the stylized form of a woman wearing a feathered hat. The elegance of the imagery was upgraded by using gold and silver on the cap in order to attract young ladies in their 20s and 30s who pursue intellectual, sophisticated, and modern styles. ● Das Gestaltungskonzept basiert auf der stilisierten Form einer Frau, die einen mit Federn geschmückten Hut trägt. Die Eleganz des Auftritts wird durch die Silber- und Goldtöne des Huts unterstützt. Zielgruppe sind gebildete, moderne Frauen zwischen 20 und 40 Jahren. ▲ Concept reposant sur une forme stylisée d'une femme portant un chapeau à plumes. Les touches dorées et argentées du chapeau renforcent cette impression d'élégance. Groupe cible: femmes sophistiquées, cultivées et modernes, entre 20 et 40 ans.

PAGE 53, *image 105* ART DIRECTOR: *Akio Okumura* DESIGNER: *Katsuji Minami* AGENCY: *Packaging Create, Inc., SIA Co., Ltd.,* CLIENT: *Cow Brand Soap, Kyoshinsha Co., Ltd.* COUNTRY: *Japan*

PAGE 54, *image 106* ART DIRECTOR: *Carol Beuttner* DESIGNERS: *Carol Beuttner, Victoria Stamm* PRODUCT PHOTOGRAPHER: *Richard Pierce* AGENCY: *Aramis Inc.* COUNTRY: *USA* ■ This packaging was intended for use as a Christmas presentation at retail department store counters. ● Verpackungsgestaltung für eine Weihnachtsaktion in Kaufhäusern. ▲ Emballage pour une promotion spéciale à l'occasion des fêtes de Noël, proposé en grandes surfaces.

PAGE 55, *images 107, 108* ART DIRECTOR: *Carol Beuttner* DESIGNERS: *Carol Beuttner, Joanne Reeves, Mike Toth, Ken Hirst* PHOTOGRAPHER: *Maria Rosledo, Product Visual Advertising* PRODUCT PHOTOGRAPHER: *Richard Pierce* CLIENT: *Tommy Hilfiger, Aramis Inc.* COUNTRY: *USA* ■ This signature-fragrance packaging was designed for the launch of collateral and gift programs. ● Die Verpackung für dieses Parfum eines Mode-Designers war für die Lancierung einer Promotions- und Geschenkaktion der Firma bestimmt. ▲ Emballage d'un parfum conçu pour le lancement de pochettes-cadeau promotionnelles.

PAGE 55, *images 109, 110* ART DIRECTOR: *Carol Beuttner* DESIGNERS: *Carol Beuttner, Victoria Stamm* PRODUCT PHOTOGRAPHER: *Richard Pierce* CLIENT: *Aramis Inc.* COUNTRY: *USA* ■ This packaging was intended for use as a Christmas presentation for retail department store counters. ● Packungsgestaltung für spezielle Weihnachtsaktionen in Kaufhäusern. (why not one caption for above and below?) ▲ Emballage pour une promotion spéciale à l'occasion des fêtes de Noël, proposé dans les grands magasins.

PAGE 56, *image 111* DESIGN DIRECTOR: *John Loring* DESIGNER: *Pierre Dinand* AGENCY: *Dinand Design* PRODUCT PHOTOGRAPHY: *Albano Ballerini* ART/PHOTO DIRECTOR: *Thuy Vuong* CLIENT: *Tiffany & Co.* ■ These bottles were designed to have a softly-sculpted look. Using bright gem colors, the design is intended to create a fluid, feminine feeling. ● Die leuchtenden Edelsteinfarben und skulpturähnlichen Flaschenformen sorgen für einen sanften, femininen Look. ▲ Les couleurs lumineuses de pierres précieuses et les formes des flacons semblables à des sculptures confèrent une impression de douceur et de féminité.

PAGE 57, *images 112, 113* ART DIRECTOR: *Kristin Breslin Sommese* DESIGNER: *Natalie Renard* PRODUCT PHOTOGRAPHY: *Dick Ackley, Penn State Photo Service* AGENCY: *Penn State School of Visual Arts (student project)* CLIENT: *Mondo dei Sogni* COUNTRY: *USA* TYPEFACE: *Handwritten* ■ Mondo dei Sogni (World of Dreams) is a gentleman's personalized grooming kit. The packaging was intended to indulge the user by creating a feeling that each product was made especially for him. ● Mondo dei Songi (die Welt der Träume) ist eine Pflegereihe für Männer. Die Packungsgestaltung soll den Eindruck vermitteln, dass jedes Produkt speziell für den Käufer gemacht wurde. ▲ Mondo dei Sogni (le Monde des Rêves) est une ligne de soins pour hommes. Le design de l'emballage doit donner l'impression que chaque produit a spécialement été créé pour le client.

PAGE 58, *images 114–17* ART DIRECTOR: *Tamotsu Yagi* PRODUCT PHOTOGRAPHER: *Steven Rahn* CLIENT: *Benetton Cosmetics USA* COUNTRY: *USA* ■ The objective was to create distinctive and innovative packaging and bottle designs. The rocket-like form of the flagship bottle was developed by combining the shape of an egg with that of a test tube. ● Die raketenähnliche Grundform der Verpackung für Benetton Cosmetics ist eine Kombination einer Eiform und der Form eines Testproduktes. Es ging um eine ungewöhnliche, innovative Verpackung für diese Kosmetiklinie. ▲ Semblable à une raquette, la forme de base de cet emballage pour Benetton est une combinaison entre la forme d'un œuf et celle d'un produit test. Il s'agissait de créer un packaging inhabituel et novateur pour cette ligne de cosmétiques.

PAGE 59, *images 118, 119* ART DIRECTOR/DESIGNER: *Elizabeth P. Ball* PRODUCT PHOTOGRAPHER: *Barry McCormick,* AGENCY: *Tom Fowler, Inc.* CLIENT: *Cheesebrough-Pond's USA Co.* COUNTRY: *USA* TYPEFACE: *Zapf Humanist* ■ Sold mainly in mass-merchandise and drug stores, this is actually two products in one package. ● Zwei Produkte in einer Verpackung, für Kaufhäuser und Drogerien konzipiert. ▲ Il s'agit de deux produits dans un seul emballage. L'ensemble est vendu principalement dans les magasins en grandes surfaces et dans les pharmacies.

PAGE 59, *images 120, 122* ART DIRECTOR: *Thuy Vuong* DESIGNERS: *Thuy Vuong, Susie Sohn* AGENCY: *Avancé Designs Inc.* CLIENT: *Tiffany & Co.* COUNTRY: *USA* ■ The objective was to design an innovative, young, fresh and stylish package that reflects the image of Tiffany and the colors displayed on the bottles. ● Hier ging es um die Gestaltung einer innovativen, jungen, frischen und eleganten Verpackung, die dem Image von Tiffany gerecht wird und die auf den Flaschen verwendeten Farben aufnimmt. ▲ Il s'agissait de créer un emballage novateur, jeune et frais, qui reflète l'image de la marque Tiffany et qui reprenne les couleurs des flacons.

PAGE 59, *image 121* ART DIRECTOR/DESIGNER: *Vanessa Eckstein* AGENCY: *Vanessa Eckstein Design* CLIENT: *Natura* COUNTRY: *Argentina* TYPEFACE: *M. Bembo, Young Baroque* ■ This design seeks to convey the company's environmental philosophy and attract the young consumer. Rather than focusing on packaging that "looks recycled," the design emphasizes the natural qualities of the product itself. ● Mit dieser Verpackung soll die umweltfreundliche Einstellung des Herstellers unterstrichen und eine junge Verbrauchergruppe angesprochen werden. Die Verpackung stellt die natürlichen Ingredienzen des Produktes in den Vordergrund und nicht das umweltfreundliche Material der Verpackung. ▲ Le concept reflète la philosophie écologique de l'entreprise et s'adresse aux jeunes consommateurs. L'accent est mis sur les qualités naturelles du produit et non pas sur l'apparence «recyclée» de l'emballage.

PAGE 59, *image 123* ART DIRECTOR: *Hans D. Flink* DESIGNERS: *Chang–Mei Lin, Suzanne Clark* AGENCY: *Hans D. Flink Design Inc.* CLIENT: *Fabergé Co.* COUNTRY: *USA* ■ These plastic barrels for a line of men's deodorants were designed with branding and color coding for various fragrances. ● Eine Linie von Deodorants für Männer, mit Farbkodierung für die verschiedenen Duftnoten. ▲ Ligne de déodorants pour hommes avec diverses couleurs pour les différentes fragrances.

PAGE 60, 61, *images 124–126* ART DIRECTOR/DESIGNER: *John Carpenter* AGENCY: *The Benchmark Group* CLIENT: *Coty/Private Portfolio* COUNTRY: *USA* PRINTER: *S.N. Burt* TYPEFACE: *Futura* PAPER: *18 pt. SBS with mylar foil* ■ The goal was to create packaging that captured the speed and spirit of the car. ● Das Rassige einer Corvette, übertragen auf die Verpackung einer Kosmetiklinie dieses Namens. ▲ L'emballage devait rendre le côté racé et dynamique d'une Corvette.

PAGE 62, *image 127* ART DIRECTOR: *Adrian Pulfer* DESIGNERS: *Adrian Pulfer, Mary Jane Callister* PHOTOGRAPHER: *Matt Mahurin* AGENCY: *Adrian Pulfer Design* CLIENT: *Raje* COUNTRY: *USA* PRINTER: *Packaging Corp. of America* TYPEFACE: *Helvetica Compressed, Sabon*

PAGE 62, *image 128* ART DIRECTOR: *Lynn Phillips* DESIGNER/ARTIST/ILLUSTRATOR: *John Evans* PHOTOGRAPHER: *John Noeding* CLIENT: *Mary Kay Inc.* COUNTRY: *USA* PRINTER: *AGI* TYPEFACE: *Kabel* ■ Positioned around the concept of aromatherapy, the design objective was to broaden the appeal of this existing line by giving it a fresh, vibrant and contemporary image. ● Neugestaltung einer auf Aroma-Therapie basierenden Produktlinie, die ein frisches, lebhaftes, zeitgemässes, neues Image brauchte. ▲ Le concept d'aromathérapie a été utilisé pour donner à cette ligne de produits une image revigorante, dynamique et contemporaine.

PAGE 63, *image 129* ART DIRECTOR: *Yoshitomo Obama* DESIGNERS: *Kazunori Umezawa,Yumiko Okaji* AGENCY/CLIENT: *Avon Products Co., Ltd.* COUNTRY: *Japan* ■ This packaging was designed as a promotional concept for powder foundation and lipstick products. ● Gestaltung einer Promotions-Verpackung für einen Fond de Teint und Lippenstift. ▲ Emballage promotionnel pour un fond de teint et des rouges à lèvres.

PAGE, 64, *image 130* ART DIRECTOR: *Alain Filiz* DESIGNERS/ARTISTS/ILLUSTRATORS: *Alain Filiz, Karen Greenberg* AGENCY: *Lonzak-Filiz* COUNTRY: *USA*

PAGE 64, *image 131* ART DIRECTOR: *Nick Ovenden* AGENCY: *Design in Action* CLIENT: *Coty* PRINTER: *Prestige Packaging* ■ This packaging and bottle design conveys retail and brand identity. ● Die Verpackungs- und Flaschengestaltung für ein Parfum. ▲ Design de l'emballage et de flacon pour un parfum.

PAGE 65, *images 132–134* CREATIVE DIRECTOR: *Amy Knapp* EXECUTIVE CREATIVE DIRECTOR: *Peter Allen* DESIGNER: *Amy Knapp* ILLUSTRATOR: *Peter Siu* CLIENT: *DFS Merchandising Ltd.* COUNTRY: *Guam* ■ This packaging was developed for Hafa Adai, a beauty care line consisting of hair and skin care products.● Packungsgestaltung für eine Haut- und Haarpflegeserie Absatzmarkt. ▲ Emballage conçu pour Hafa Adi, une ligne de produits de soins corporels et capillaires .

PAGE 66, *image 135* ART DIRECTOR: *Mary Lewis* DESIGNER: *Lucilla Scrimgeour* PRODUCT PHOTOGRAPHER: *Robin Broadbent* AGENCY: *Lewis Moberly* CLIENT: *Next plc.* COUNTRY: *UK*

PAGE 67, *images 136, 138* ART DIRECTORS: *Thomas Fairclough, Tom Antista* DESIGNER: *Thomas Fairclough* ARTIST/ILLUSTRATOR: *Flatland* PHOTOGRAPHER: *Jody Dole, Michael West Photography* AGENCY: *Antista Fairclough Design* CLIENT: *Montsource, Inc.* COUNTRY: *USA* ■ This design utilizes a combination of classical sculpture, contemporary shapes and textures to create a strong masculine identity. ● Um ein starkes männliches Image zu etablieren, wurden hier Elemente der klassischen Bildhauerei mit modernen Formen kombiniert. ▲ Combinaison entre sculpture classique, formes et textures modernes afin de créer une forte identité masculine.

PAGE 67, *image 137* ART DIRECTOR: *Sigi Mayer* AGENCY: *Pharmaperformance GmbH* CLIENT: *Simons GmbH* COUNTRY: *Germany* PAPER: *Mikro Well* ■ This package was created to contain pills related to the health of hair and nails. ● Verpackung für Pillen zur Stärkung von Haar und Fingernägeln. ▲ Emballage pour des pilules destinées à fortifier les ongles et les cheveux

PAGE 68, *image 139, 140* CREATIVE DIRECTOR: *Phyllis Aragaki* DESIGNER: *Frances Ullenberg* AGENCY: *Desgrippes Gobé & Associates* CLIENT: *Victoria's Secret* ■ The objective was to create packaging that was modern, sensual, and sophisticated with a hint of mystery. ● Hier ging es um die Gestaltung einer Verpackung, die modern, sinnlich, anspruchsvoll und auch ein bisschen geheimnisvoll wirken sollte. ▲ Il s'agissait de créer un emballage moderne, sensuel et sophistiqué, avec un soupçon de mystère.

PAGE 70, *images 141, 142* ART DIRECTOR: *Fabio Ongarato* DESIGNERS: *Leonard Hamersfeld, Tim Richardson* PRODUCT PHOTOGRAPHER: *Greg Delves* AGENCY: *Fabio Ongarato Design* CLIENT: *Kusco-Murphy Salon Pty. Ltd.* COUNTRY: *Australia* TYPEFACE: *Gill Sans, Meta* ■ The prerequisite was to produce an environmental package compatible with the 100% vegetarian product. Aluminum, glass, and recycled rubber stoppers were utilized. ● Für diese Verpackung wurden Aluminium, Glas und wiederverwertete Gummi-Verschlüsse verwendet, um dem umweltfreundlichen Anpruch eines hundertprozentig vegetarischen Produktes gerecht zu werden. ▲ De l'aluminium, du verre et des bouchons en caoutchouc recyclés ont été utilisés pour cet emballage écologique contenant un produit 100% naturel.

PAGE 71, *image 143, 144* ART DIRECTOR: *Claus Koch* DESIGNERS: *Peter Engelbardt,Claus Koch* PHOTOGRAPHER: *Packshot Boys* AGENCY: *Claus Koch Corporate Communications* CLIENT: *Wella AG* COUNTRY: *Germany* TYPEFACE: *Futura condensed, regular and half bold* PAPER: *semi-mat tracing foil (transparent paper)* ■ Complete redesign of a product range for a worldwide relaunch of "System Professional", which claims to be a high-quality, cosmetic product. ● Vollständige Neugestaltung für einen weltweiten Relaunch der Haarkosmetikserie «System Professional» als hochwertiges kosmetisches Produkt für Coiffeur-Salons. ▲ Nouvelle conception pour un nouveau lancement international de la ligne de produits de soins capillaires «System Professional» ultra-performants.

PAGE 71, *images 145, 146* ART DIRECTOR/DESIGNER: *Angelo Sganzerla* ARTIST/ILLUSTRATOR: *Alfonso Goi* AGENCY: *Angelo Sganzerla* CLIENT: *L'Erbaolario* ■ This packaging was created to emphasize the qualities of the client's naturally derived shampoos, perfumes, toiletries, and body care products. ● Bei der Verpackung ging es um die Betonung der natürlich Ingredienzen der Shampoos, Parfums und Körperpflegemittel. ▲ L'emballage met l'accent sur les qualités des produits fabriqués naturellement: shampooings, parfums et produits pour les soins corporels.

PAGE 71, *image 147* ART DIRECTOR/DESIGNER: *Nick Ovenden* AGENCY: *Design in Action* CLIENT: *Coty* COUNTRY: *United Kingdom* PRINTER: *Prestige Packaging* ■ The design objective was to establish brand identity, packaging and point-of-purchase material for one of the first unisex fragrances .● Ein starker

Markenauftritt, unterstützt durch die Verpackung und Display-Material für eines der ersten Unisex-Parfums, das für den Massenmarkt bestimmt ist. ▲ Une forte identité de marque renforcée par l'emballage et le matériel de présentation pour l'un des premiers parfums unisexes destiné à la grande distribution.

PAGE 71, *image 148* ART DIRECTOR: *Joël Desgrippes* DESIGNER: *Corinne Restrepo* PRODUCT PHOTOGRAPHER: *Ch. Gros* AGENCY: *Degrippes Gobe & Associates* CLIENT: *Boucheron* COUNTRY: *France* ■ Jaïpur is the residence of the Maharajah and mythical location where goldsmiths craft lucky-charm bracelets for newlyweds. The bracelet bottle reflects this, combining rock crystal, polished glass, gold, and sapphire blue tints. ● Diese Flasche mit einer Kombination von Kristall, geschliffenem Glas sowie Gold- und saphirblauen Tönen ist eine Interpretation dessen, was ihr Name impliziert: Jaïpur, Hauptstadt von Rajasthan, Indien, Sitz des Maharadschas und mystischer Ort, wo Goldschmiede Armbänder als Glücksbringer für Neuvermählte anfertigen. ▲ Jaïpur est la résidence du maharajah et l'endroit mythique où les orfèvres créent des bracelets porte-bonheur pour les jeunes mariés. Le flacon aux teintes or et saphir alliant cristal et verre poli est à l'image du nom du parfum, riche en évocations.

PAGE 71, *image 149* CREATIVE DIRECTOR: *Phyllis Aragaki* DESIGNER: *Marion Clédat* PRODUCT PHOTOGRAPHER: *Philippe Houzé* AGENCY: *Desgrippes Gobé & Associates* CLIENT: *Victoria's Secret* COUNTRY: *USA* ■ The goal of the packaging for this product line was to communicate the vanilla flavor and a sense of the exotic—as if the fragrances had been collected from travels to the Far East. The colors were inspired by that region as well as by the individual flavors. ● Die Verpackung dieser Linie evoziert den Vanilleduft und die Exotik der Produkte – als seien die Duftstoffe bei einer Reise in den Fernen Osten gesammelt worden. Die Farben der ganzen Linie sind eine Anspielung auf diese Region sowie auf die verschiedenen Duftnoten. ▲ L'emballage de cette ligne de produits évoque le parfum de vanille et l'exotisme – comme si les fragrances avaient été rassemblées lors de voyages en Extrême-Orient. Les couleurs utilisées pour cette ligne ainsi que les différentes fragrances s'inspirent de cette région du monde.

PAGE 71, *image 150* CREATIVE DIRECTOR: *Phyllis Aragaki* DESIGNER: *Anne Swan* PRODUCT PHOTOGRAPHER: *Philippe Houzé* ARTIST/ILLUSTRATOR: *Marion Clédat* AGENCY: *Desgrippes Gobé & Associates* CLIENT: *Victoria's Secret* COUNTRY: *USA* ■ The goal was to create a unique and proprietary fruit-based toiletries line within the client's established graphic language. The line was differentiated through use of sensual and soft fruit icon illustrations. ● Packungsgestaltung für eine Körperpflegelinie auf Fruchtbasis. Aufbauend auf dem bereits gut etablierten graphischen Auftritt der Firma, wurden sinnliche, sanfte Darstellungen von Früchten eingesetzt, um dieser Linie ihr spezielles Gesicht zu geben. ▲ Il s'agissait de créer une ligne de produits de soins uniques à base de fruits en reprenant l'identité graphique déjà bien établie du client. Les illustrations de fruits, empreintes de sensualité et de douceur, ont contribué à créer un effet spécial.

PAGE 72, 73; *images 151–157* ART DIRECTOR: *Jaimie Alexander* DESIGNERS: *Jaimie Alexander, Mandy Putnam, Paul Westrick* PHOTOGRAPHER: *Mark Steele* AGENCY: *Fitch Inc.* CLIENT: *Hush Puppies Company* COUNTRY: *USA* TYPEFACE: *Copperplate, Hand-drawn* PAPER: *Clay-coated News* ■ The designers sought to create brand positioning for Hush Puppies children's line, and to develop the appropriate communication components, including labels, hangtags, packaging, retail presentation, and sales materials. ● Hier ging es um die Markenetablierung und den gesamten Auftritt, einschliesslich Etiketten, Anhängern, Verpackung, Ladendisplays, Verkaufs- und Kommunikationsmaterial für eine Hush-Puppies-Marke für Kinder. ▲ Il s'agissait de positionner la marque pour enfants Hush Puppies en concevant des étiquettes, des emballages, des présentoirs et des dépliants distribués dans les différents points de vente.

PAGE 74, *image 158* ART DIRECTORS: *Maruchi Santana, Millie Hsi* DESIGNER: *Millie Hsi* PHOTOGRAPHER: *Bart Gorin* AGENCY: *Parham Santana, Inc.* CLIENT: *Via International Group* ■ This packaging was designed to attract a newer consumer without alienating the die hard Harley owner. ● Mit dieser Verpackung sollten neue Käufer angezogen werden, ohne den eingefleischten Harley-Davidson-Besitzer vor den Kopf zu stossen. ▲ Cet emballage avait pour objectif d'attirer de nouveaux consommateurs sans s'attirer pour autant les foudres des mordus des produits Harley-Davidson.

PAGE 75, *image 159* ART DIRECTORS: *Maruchi Santana, Millie Hsi* DESIGNER: *Millie Hsi* DESIGNER: *Lori Ann Reinig* PHOTOGRAPHER: *Bart Gorin* AGENCY: *Parham Santana, Inc.* CLIENT: *Via International Group Fairfield, NJ* ■ This set of four collectable tins served as a point-of-sale display and helped distinguish the client in a mature and crowded eyewear market. ● Dieser Satz von vier Sammler-Dosen diente als Laden-Display und sollte dazu beitragen, das Produkt von der Masse im riesigen Brillenmarkt abzuheben. ▲ Cet ensemble de quatre boîtes en fer blanc a servi de présentoir de magasin et avait pour objectif de démarquer clairement les produits du client.

PAGE 76, *image 160* ART DIRECTION: *Sazaby Graphic Design* DESIGNER: *Chie Kusakari* CLIENT: *Sazaby Inc.* COUNTRY: *Japan* ■ An acrylic plate and silver bolts are used as artificial materials to complement this silver metallic watch. Pink, corrugated cardboard paper is piled up for warmth and brightness. Maintaining the unisex design of the product, this package is meant to be not too hard, but not too sweet. ● *Akryl und Silberschrauben wurden für diese Silbermetallic-Uhr verwendet. Rosafarbener Wellkarton sorgt für Wärme und Farbigkeit. Da das Produkt für Männer und Frauen bestimmt ist, sollte der Auftritt weder zu hart noch zu süsslich wirken.* ▲ *De l'acryle et des boulons en argent ont été utilisés pour cette montre métallique. Le carton ondulé rose donne une impression de chaleur et de luminosité. Le produit s'adressant aussi bien aux femmes qu'aux hommes, il s'agissait de créer un concept unisexe, ni trop doux ni trop dur.*

PAGE 77, *image 161* ART DIRECTOR: *Alan Chan* DESIGNERS: *Alan Chan, Oscar Tse* ARTISTS/ILLUSTRATORS: *Oscar Tse, Chow Hang Fai* AGENCY: *Alan Chan Design Company* CLIENT: *Swank Inspiration* COUNTRY: *Hong Kong* ■ *The theme of this range of packaging was developed and based upon the five elements of nature found in Chinese philosophy: Metal, Wood, Water, Fire, and Earth. The butterfly motif was used to symbolize the beauty of the fashion world.* ● *Thema dieser Verpackungslinie sind die fünf Naturelemente der chinesischen Philosophie: Metall, Holz, Wasser, Feuer, Erde. Das Schmetterlingsmotiv symbolisiert die Schönheit der Modewelt.* ▲ *Le thème de ces emballages repose sur les cinq éléments de la nature que l'on trouve dans la philosophie chinoise: le Métal, le Bois, l'Eau, le Feu et la Terre. Le motif de papillons symbolise la beauté inhérente au monde de la mode.*

PAGE 78, *image 162* CREATIVE DIRECTOR/DESIGNER: *Amy Knapp* EXECUTIVE CREATIVE DIRECTOR: *Peter Allen* ARTIST/ILLUSTRATOR: *Gary Baseman* CLIENT: *DFS Merchandising Ltd. Countries: Asia & US* TYPEFACE: *Hand-drawn, brush stroke calligraphy* PAPER: *Speckletone and various others (recycled on collateral pieces)* ■ *The logo depicts the idea of a global village and children of all races living together on this planet. The design includes a shop concept for children's departments in DFS stores worldwide.* ● *Das Logo basiert auf der Idee eines globalen Dorfes, in dem Kinder aller Rassen zusammenleben. Das Konzept wird weltweit in den Kinderabteilungen der DFS-Läden angewendet.* ▲ *Le logo se base sur l'idée d'un village multiracial où vivent des enfants du monde entier. Ce concept international est utilisé dans tous les rayons enfants des magasins DFS*

PAGE 78, 79; *images 163, 164* ART DIRECTOR: *Kit Hinrichs* DESIGNER: *Jackie Foshaug* AGENCY: *Pentagram Design* CLIENT: *The Gymboree Corporation* ■ *This package design project includes the design and implementation of a new visual identity for the client. The "new look" replaces traditional primary colors* ● *Packungsgestaltung im Rahmen eines Erscheinungsbildes des Kunden. Zum «neuen Look» gehört eine neue Farbpalette, die die traditionellen Grundfarben ersetzt.* ▲ *Emballage conçu dans le cadre de l'identité visuelle du client. Pour ce «look» revisité, de nouveaux coloris remplacent les traditionnelles couleurs primaires.*

PAGE 80, *image 165* ART DIRECTOR: *Doo H. Kim* DESIGNERS: *Dongil Lee, Seunghee Lee* AGENCY: *DooKim Design* CLIENT: *Trugen, Nasan Industrial Co. Ltd.* COUNTRY: *Korea* ■ *The shape of the lion was designed to represent a modernized, refined image and to symbolize European high society.* ● *Der Löwe dient hier als Ausdruck eines moderneren, anspruchsvollen Produktes und als Symbol für die europäische High-Society.* ▲ *Le lion donne une image moderne, raffinée, et symbolise en même temps la haute société européenne.*

PAGE 80, *image 166* ART DIRECTOR: *Sharon Werner* DESIGNERS: *Sharon Werner, Sarah Nelson* PRODUCT PHOTOGRAPHER: *Darrell Eager* AGENCY: *Werner Design Werks* CLIENT: *HMI Resources* COUNTRY: *USA* ■ *This packaging system for textiles imported from Laos was designed to accent the scarves' unique and delicate beauty. The book tells how the scarves are made and their special history. The box itself is a sales kit to go to retail buyers.* ● *Diese Verpackung sollte die Schönheit und Feinheit der aus Laos importierten Tücher zum Ausdruck bringen. Eine Broschüre informiert über Herstellung und Geschichte der Tücher. Die Box selbst ist für den Einzelhandel bestimmt.* ▲ *Le packaging devait mettre en valeur la beauté et la finesse de foulards importés du Laos. Une brochure adjointe informe sur la fabrication et l'histoire des foulards. La boîte est destinée au commerce de détail.*

PAGE 80, *image 167* CREATIVE DIRECTORS: *Wade Koniakowsky, Jon Gothold* ART DIRECTOR/DESIGNER/ILLUSTRATOR: *Jeff Labbé* PHOTOGRAPHER: *Kimball Hall* AGENCY: *dGWB Advertising* CLIENT: *Vans Shoes Inc.* COUNTRY: *USA* PRINTER: *Scope Packaging* TYPEFACE: *Courier Bold, Regular* PAPER: *French Paper Company* ■ *This packaging was designed to reflect a sense of summer and also to serve as a point-of-purchase promotion.* ● *Mit dieser Verpackung sollte das Gefühl von Sommer zum Ausdruck kommen. Sie diente auch als Laden-Promotion.* ▲ *Cet emballage devait être à l'image de l'été.*

PAGE 80, *image 168* DESIGNER: *Jason O'Hara* PHOTOGRAPHER: *Melanie Bridge* AGENCY: *BNA Design* CLIENT: *Manawatu Knitting Mills* COUNTRY: *New Zealand* PRINTER: *Wellington Print Finishers Ltd.* TYPEFACE: *Stempel Garamond & Shelley Allegro* PAPER: *Gainsborough* ■ *This packaging was designed as point-of-sales material for the Lyle & Scott range of clothing.* ● *Gestaltung der Ladenpromotion für die Bekleidungsmarke Lyle & Scott.* ▲ *Cet emballage a été conçu pour être mis en valeur dans les points de vente de la ligne de vêtements Lyle & Scott.*

PAGE 81, *image 169* ART DIRECTOR/DESIGNER/ILLUSTRATOR: *Linda Fountain* AGENCY: *Linda Fountain Design* CLIENT: *Strung Out Jewelry* COUNTRY: *USA* ■ *These boxes were produced for a small jewelry company. Each box was hand-decorated to coordinate with the color and style of the jewelry inside.* ● *Jede dieser Schachtein einer kleinen Schmuckfirma wurde von Hand verziert, abgestimmt*

auf Farbe und Stil ihres Inhalts. ▲ *Chacune de ces petites boîtes d'une bijouterie a été décorée à la main en fonction de la couleur et du style du contenu.*

PAGE 82, *image 170* ART DIRECTOR: *Alan Disparte* DESIGNERS: *Alan Disparte, Lucia Matioli* AGENCY: *The Gap, Inc.* CLIENT: *Old Navy Clothing Co.* COUNTRY: *USA* TYPEFACE: *Franklin Gothic, Coronet* PAPER: *Label Stock* ■ *This packaging was created for food products sold in Old Navy's New York coffee shop.* ● *Verpackungsgestaltung für Lebensmittel, die im Old Navy's-New-York-Café verkauft werden.* ▲ *Emballage créé pour des produits alimentaires vendus dans le café Old Navy à New York.*

PAGE 83, *image 171* CREATIVE DIRECTOR/DESIGNER: *Amy Knapp* EXECUTIVE CREATIVE DIRECTOR: *Peter Allen* PRODUCT PHOTOGRAPHER: *Deborah Jones* ILLUSTRATOR: *Ann Field* CLIENT: *DFS Merchandising Ltd.* COUNTRY: *USA* PRINTER: *Mid-Cities* ■ *This packaging is designed to capture the spirit of California through bright colors and whimsical illustrations depicting the carefree California lifestyle.* ● *Mit den leuchtenden Farben und originellen Illustrationen dieser Verpackungsgestaltung für ein Café sollte der sorglose kalifornische Lebensstil zum Ausdruck kommen.* ▲ *Les couleurs vives et les illustrations originales de cet emballage de café sont à l'image du style de vie insouciant des Californiens.*

PAGE 83, *image 172* ART DIRECTOR/DESIGNER: *Yasuo Tanaka* AGENCY: *Package Land Company, Ltd.* CLIENT: *Arab Coffee Company, Ltd.* COUNTRY: *Japan* ■ *Corporate logo mark and coffee package design* ● *Marken- und Verpackungsgestaltung für einen Kaffee.* ▲ *Conception du logo d'entreprise et de l'emballage d'un café.*

PAGE 83, *image 173* ART DIRECTOR: *Steve Coleman* DESIGNER: *Chris De Lisen* AGENCY: *Elton Ward Design* COUNTRY: *Australia* ■ *This is a short-run production designed to be filled with coffee as a client gift.* ● *Eine spezielle Verpackung für Kaffee, der den Kunden als Geschenk übergeben wurde.* ▲ *Emballage spécial pour du café, remis en cadeau au client.*

PAGE 84, *image 174* ART DIRECTOR: *David Ceradini* DESIGNER: *Joe Cuticone* PRODUCT PHOTOGRAPHER: *David Lazarus* ARTIST/ILLUSTRATOR: *Tim Shannon* AGENCY: *Ceradini Design* CLIENT: *Compass Foods* ■ *The idea was to create indulgent packaging for instant cappuccino.* ● *Hier ging es um eine genussversprechende Verpackung für sofortlöslichen Cappuccino.* ▲ *Il s'agissait de créer un emballage évoquant la richesse de l'arôme d'un cappuccino instantané.*

PAGE 84, *image 175* ART DIRECTOR/DESIGNER: *David Ceradini* ARTIST/ILLUSTRATOR: *David Watts Boone,* AGENCY: *Ceradini Design* CLIENT: *Compass Foods* ■ *The design promotes premium whole bean coffees from around the world.* ● *Packungsgestaltung für ungemahlenen Kaffee erster Güte aus verschiedenen Anbaugebieten.* ▲ *Emballage pour du café en grains de qualité supérieure cultivé dans le monde entier.*

PAGE 84, *image 176* ART DIRECTOR: *Jack Anderson* DESIGNERS: *Jack Anderson, Julie Lock, Julie Keenan* PRODUCT PHOTOGRAPHER: *Tom McMackin* ARTIST/ILLUSTRATOR: *John Fretz* AGENCY: *Hornall Anderson Design Works, Inc.* CLIENT: *Starbucks Coffee Company* COUNTRY: *USA* PRINTER: *Riverwild Internationall (carton), Anchor Glass (bottle)* TYPEFACE: *Reckleman, Copperplate, customized hand-lettering* PAPER: *Kraft, glass* ■ *This customized bottle shape and four-pack carton was created for a new sparkling coffee drink.* ● *Flaschen- und Kartongestaltung für ein Erfrischungsgetränk auf Kaffee-Basis.* ▲ *Conception des bouteilles et du pack de quatre pour une boisson rafraîchissante à base de café.*

PAGE 85, *images 177, 178* ART DIRECTOR/DESIGNER: *Jeff Compton* PRODUCT PHOTOGRAPHER: *Geoffrey Sokol* AGENCY: *Weiss, Whitten, Stagliano* CLIENT: *Bucks County Coffee* COUNTRY: *USA* TYPEFACE: *Spectrum* ■ *This classic design was meant to apppeal to a wide audience. The objective was to balance the down-home feeling of Buck County, Pennsylvania with the modern ethos of specialty coffee.* ● *Dieses klassische Design sollte eine breite Konsumentengruppe ansprechen. Einerseits ging es um das 'wie-zu-Hause'-Gefühl, andererseits um das moderne Image einer anspruchsvollen Spezialmischung.* ▲ *Design classique s'adressant à un large public. L'objectif était de donner l'impression de sentir «comme chez soi» et de transmettre en même temps une image moderne pour un mélange de café de premier choix.*

PAGE 86, *images 179, 181* ART DIRECTOR/DESIGNER/ARTIST/ILLUSTRATOR: *Neil Powell* PHOTOGRAPHER: *Karen Capucilli* AGENCY: *Duffy Design* CLIENT: *The Coca-Cola Company* COUNTRY: *USA* ■ *This brand identity and package system for a line of 100% juices and juice blends is intended to communicate attributes of an honest, straight-forward product.* ● *Packungsfamilie für reine Säfte und Saftgetränke. Hier ging es um die Darstellung eines ehrlichen, soliden Produktes.* ▲ *Famille d'emballages pour des jus de fruits 100% naturels et des cocktails de jus de fruits. Simplicité et honnêteté, tels étaient les deux critères qui devaient transparaître à travers l'image de ces produits.*

PAGES 86, 87; *images 180, 183* ART DIRECTOR/DESIGNER/ARTIST/ILLUSTRATOR: *Neil Powell* PHOTOGRAPHER: *Mark La Favor* AGENCY: *Duffy Design* CLIENT: *The Coca-Cola Company* COUNTRY: *USA* ■ *This package design was created to balance a sense of tradition with a more contemporary and natural look.* ● *Aufgabe dieser Verpackung war die Verbindung von Tradition mit einem modernen, natürlichen Look.* ▲ *Le design de cet emballage devait allier tradition, nature et modernisme.*

PAGE 86, *image 182* ART DIRECTOR/DESIGNER/ARTIST/ILLUSTRATOR: *Neil Powell* PHOTOGRAPHER: *Karen Capucilli* AGENCY: *Duffy Design* CLIENT: *The Coca-Cola Company* COUNTRY: *USA* ■ *This is an upscale product with a higher price point. The design intends to reposition this product as the best orange juice one can buy.* ● *Hier ging es um die Repositionierung eines Produktes der gehobenen Preisklasse als der beste Orangensaft, den man kaufen kann.* ▲ *Repositionnement d'un produit de premier choix au prix élevé. Il s'agissait d'imposer ce jus d'oranges comme le meilleur que l'on puisse acheter sur le marché.*

PAGE 88, *image 184* ART DIRECTOR: *John Blackburn* DESIGNER: *Belinda Duggan* ARTIST/ILLUSTRATOR: *John Geary* AGENCY: *Blackburn's Limited* CLIENT: *MD Foods plc.* COUNTRY: *United Kingdom* PRINTER: *Tetrapack* TYPEFACE: *Gill Sans* ■ *The packaging design gives strong branding through simple emotive graphics. The product is also separated from competitors as the bold colors make it distinctive in a market dominated by white packs.* ● *Die Packung verdankt ihre Wirkung der Illustration und den kräftigen Farben, was das Produkt von den üblicherweise weissen Milchverpackungen deutlich unterscheidet.* ▲ *L'originalité de cet emballage est due à son illustration et aux couleurs soutenues, lesquelles doivent permettre de démarquer ce packaging des traditionnels emballages blancs.*

PAGE 88, *image 185* ART DIRECTOR: *Dimitri Karavias* DESIGNERS: *Dimitri Karavias, Liz Knight* AGENCY: *Minale, Tattersfield + Partners, Ltd.* CLIENT: *Kuwait Danish Dairy Company* COUNTRY: *Kuwait* PRINTER: *Tera Pak* TYPEFACE: *Arabic Calligraphy, Gill, Bembo* PAPER: *Standard Tetrapak carton paper* ■ *The design for this packaging employs the pattern of an Arab carpet to create a traditional image. The color of the type is intended to denote green pastures.* ● *Mit dem arabischen Teppichmuster dieser Packung soll das Gefühl von Tradition erzeugt werden, während die Schrift auf grüne Weiden hinweist.* ▲ *Le motif d'un tapis arabe sur cet emballage évoque l'idée de tradition, tandis que la couleur des caractères est sensée faire penser à de vertes prairies.*

PAGE 88, *image 186* CREATIVE DIRECTOR: *Ian Tudhope* DESIGNER: *Cathy Russell* PHOTOGRAPHER: *Creeshla Hewitt* ILLUSTRATOR: *Damian Glass* AGENCY: *Tudhope Associates* CLIENT: *Ault Foods Limited* COUNTRY: *Canada* ■ *The packaging was intended to communicate an image of purity and freshness while giving the product a premium positioning.* ● *Diese Verpackung soll das Gefühl von Reinheit und Frische eines qualitativ hochwertigen Produktes erwecken.* ▲ *Cet emballage doit transmettre une image de pureté et de fraîcheur tout en soulignant la qualité du produit.*

PAGE 88, *image 187* ART DIRECTOR: *John Hornall* DESIGNERS: *John Hornall, Jana Nishi, Debra Hampton, Leo Raymundo, Mary Chin Hutchinson* PRODUCT PHOTOGRAPHER: *Tom McMackin* ARTIST/ILLUSTRATOR: *Carolyn Vibbert* AGENCY: *Hornall Anderson Design Works, Inc.* CLIENT: *Darigold* COUNTRY: *USA* PRINTER: *James River* TYPEFACE: *Futura* ■ *The packaging was created to update the product's current look.* ● *Mit dieser Verpackung soll der Auftritt des Produktes verjüngt werden.* ▲ *Cet emballage devait rajeunir l'image du produit.*

PAGE 88, *image 188* ART DIRECTOR: *Zeljko Borcic* DESIGNER: *Tamara Reljan-Musulin* ARTIST/ILLUSTRATOR: *Masa Sirovica* AGENCY: *Graffiti Design* CLIENT: *Dukat Dairy of Zagreb, Croatia* PRINTER: *Elopak Trading AG, Swiss* TYPEFACE: *Various* PAPER: *Polyehylene-coated cardboard* ■ *This packaging design was created for a major producer of Croatian milk products.* ● *Packungsgestaltung für den grössten Hersteller von Milchprodukten in Kroatien.* ▲ *Le design de cet emballage a été créé pour un important producteur de produits laitiers croates.*

PAGE 89, *image 189* ART DIRECTOR: *John Ball* DESIGNERS: *John Ball, Mike Brower* ARTIST/ILLUSTRATOR: *Tracy Sabin* AGENCY: *Mires Design Inc.* CLIENT: *Ken C. Smith Advertising* COUNTRY: *USA* ■ *The clean design and cool color scheme for this fat-free, cholesterol-free product is intended to give the consumer a feeling that he or she is consuming a healthier product.* ● *Das klare Design und die kühlen Farben sollen dem Verbraucher das Gefühl geben, dass es sich bei dieser fett- und cholesterinfreien Milch um ein gesundes Produkt handelt.* ▲ *La pureté du design et la fraîcheur des coloris doivent donner l'impression au consommateur que ce produit diététique sans cholestérol est un produit sain.*

PAGE 89, *image 190* ART DIRECTOR: *Mike Thomas* DESIGNERS: *Sharon Werner, Todd Bartz* PRODUCT PHOTOGRAPHER: *Darrell Eager* ARTIST/ILLUSTRATOR: *Sharon Werner* AGENCY: *Werner Design Werks* CLIENT: *Target Stores* COUNTRY: *USA* ■ *The goal was to create a packaging system that is fresh, clean, and versatile enough to be recognized by the consumer, yet still fits with a variety of products.* ● *Diese Verpackungsgestaltung sollte beim Konsumenten einen nachhaltigen Eindruck von Frische und Sauberkeit hinterlassen und sich dabei auch für weitere Produkte eignen.* ▲ *Cet emballage devait donner une impression de fraîcheur et de propreté et être également utilisable pour d'autres produits.*

PAGE 90, *image 191* CREATIVE DIRECTOR: *Kan Tai-keung* ART DIRECTORS: *Freeman Lau Siu Hong, Eddy Yu Chi Kong* DESIGNERS: *Freeman Lau Siu Hong, Eddy Yu Chi Kong, Joyce Ho Ngai Sing* COMPUTER GRAPHICS: *Benson Kwun Tin Yau* AGENCY: *Kan Tain-keung Design & Associates* CLIENT: *Effem Foods Inc.* COUNTRY: *USA* ■ *This packaging was created for the Chinese New Year.* ● *Für das chinesische Neujahr entworfene Packung.* ▲ *Emballage créé pour le Nouvel An chinois.*

PAGE 91, *image 192, 193* ART DIRECTOR/DESIGNER: *Angelo Sganzerla* ARTIST/ILLUSTRATOR: *Alfonso Goi* AGENCY: *Angelo Sganzerla* CLIENT: *Andrea Stainer*

PAGE 91, *image 194* CREATIVE DIRECTOR/DESIGNER: *Amy Knapp* EXECUTIVE CREATIVE DIRECTOR: *Peter Allen* PRODUCT PHOTOGRAPHER: *Deborah Jones* ILLUSTRATOR: *Ann Field* CLIENT: *DFS Merchandising Ltd.* COUNTRY: *USA* PRINTER: *Mid-Cities* ■ *This food product packaging is designed to capture the spirit of California through bright colors and whimsical illustrations depicting the carefree California lifestyle.* ● *Leuchtende Farben und verrückte Illustrationen als Interpretation des sorglosen kalifornischen Lebensstils kennzeichnen die Verpackungen der von einem Café angebotenen Lebensmittel.* ▲ *Couleurs vives et illustrations détonantes, une interprétation du style de vie insouciant des Californiens transposée sur cet emballage de produits alimentaires.*

PAGE 91, *image 195* CREATIVE DIRECTOR/DESIGNER: *Amy Knapp* EXECUTIVE CREATIVE DIRECTOR: *Peter Allen* PRODUCT PHOTOGRAPHER: *Deborah Jones* ILLUSTRATOR: *Ann Field* CLIENT: *DFS Merchandising Ltd.* COUNTRY: *Italy* PRINTER: *Bianchi* ■ *This packaging for Italian cookies was created to be sold in the Pacific Rim countries.* ● *Packungsgestaltung für italienisches Gebäck, das in die Pazifikländer ausgeführt wird.* ▲ *Emballage pour des biscuits italiens destiné aux pays du Pacifique.*

PAGE 91, *image 196* ART DIRECTOR: *Mark Sackett* DESIGNERS: *Mark Sackett, Wayne Sakamoto, Clark Richardson* DESIGNER FIRM: *Sackett Design Associates* CLIENT: *DFS Group Limited* TYPEFACE: *Bernard Modern* ■ *This packaging was intended to give a sense of the history behind the game of Chinese chess.* ● *Diese Verpackung für chinesisches Gebäck ist eine Anspielung auf das chinesische Schachspiel und dessen Tradition.* ▲ *Emballage évoquant le jeu d'échecs chinois et sa tradition.*

PAGE 91, *image 197* ART DIRECTOR: *Mark Sackett* DESIGNERS: *Mark Sackett, Wayne Sakamoto, Clark Richardson* DESIGN FIRM: *Sackett Design Associates* CLIENT: *DFS Group Limited* TYPEFACE: *Bernard Modern* ■ *The packaging this product centers around the Taiwan Pagodas, one of Taiwan's most popular tourist attractions.* ● *Thema dieser Verpackung sind die taiwanischen Pagoden, die zu den grössten Touristenattraktionen des Landes gehören.* ▲ *Thème de cet emballage: les pagodes de Taiwan, une des attractions touristiques les plus populaires du pays.*

PAGE 92, *image 198* ART DIRECTOR/DESIGNER: *Kate Fujishima* AGENCY: *Murrie Lienhart Rysner and Associates* CLIENT: *Dayton-Hudson* COUNTRY: *USA* TYPEFACE: *Hand-lettering, Ruzicka* ■ *This packaging was designed to position a new mint candy as a convenient, self-indulgent snack.* ● *Packungsgestaltung für die Einführung eines neuen, handlichen Snacks mit Pfefferminze zum Verwöhnen.* ▲ *Emballage pour un nouveau bonbon à la menthe.*

PAGE 92, *image 199* ART DIRECTOR/DESIGNER/ILLUSTRATOR: *Laura C. Glauser* CREATIVE DIRECTOR: *Pactrick Leo Huttenmoser* PRODUCT PHOTOGRAPHER: *Claude Joray* AGENCY: *Atelier Zone* CLIENT: *OS3, Organisation für Fairen Handel* COUNTRY: *Switzerland* PRINTER: *Wittwer Offset AG* TYPEFACE: *Futura Book, Futura Condensed, Bodoni* PAPER: *Ikonofix, 90g, white, SK3, glazed, coated on both sides* ■ *This redesign of the packaging for a range of chocolates was created for an organization committed to paying fair prices to the suppliers of raw materials.* ● *Überarbeitung des Auftritts für sechs Sorten von Schokoladen des oberen Preissegments Der Auftraggeber ist eine Organization, die sich zur Zahlung von fairen Preisen an die Rohstofflieferanten verpflichtet hat.* ▲ *Nouvel emballage pour un assortment de chocolats surfins. Le client est une organization réputé pour payer des prix honnêtes à ses fournisseurs de matieres brutes.*

PAGE 93, *image 200* DESIGNERS: *Kenzo Nakagawa, Nobuyane Hiroyasu, Norikami Satoshi* CLIENT: *Tower Shop* AGENCY: *NDC Graphics* COUNTRY: *Japan* PRINTER: *Tokyo Shiko Co., Ltd.* TYPEFACE: *Rotis Semi Sans* PAPER: *Coated paper 110g.* ■ *This packaging was designed as a souvenir of the 296-meter-tall Landmark Tower in Yokohama.* ● *Souvenirverpackung in der Form eines Wahrzeichens von Yokohama, dem 296m hohen Turm.* ▲ *Emballage-souvenir de la Tour de Yokohama d'une hauteur de 296 mètres.*

PAGE 93, *images 201–203* ART DIRECTOR: *Kenzo Nakagawa* DESIGNERS: *Kenzo Nakagawa, Nobuyane Hiroyasu, Norikami Satoshi* ILLUSTRATOR/LETTERING *Ajioka Shintaro* CLIENT: *Tower Shop* AGENCY: *NDC Graphics* COUNTRY: *Japan* PRINTER: *Tokyo Shiko Co., Ltd.* TYPEFACE: *Rotis Semi Sans* PAPER: *Coated 110g.* ■ *This packaging was designed as a souvenir of Yokohama.* ● *Verpackung als Souvenir von Yokohama.* ▲ *Emballage-souvenir de Yokohama.*

PAGE 94, *image 204* CREATIVE DIRECTOR: *Kan Tai-keung* ART DIRECTORS: *Freeman Lau Siu Hong, Eddy Yu Chi Kong* DESIGNERS: *Freeman Lau Siu Hong, Eddy Yu Chi Kong, Joyce Ho Ngai Sing* COMPUTER GRAPHICS: *Benson Kwun Tin Yau* AGENCY: *Kan Tain-keung Design & Associates* CLIENT: *Effem Foods Inc.* COUNTRY: *USA* ■ *This packaging was created for the Chinese New Year* ● *Diese Verpackung wurde speziell für das chinesische Neujahr entworfen.* ▲ *Emballage créé pour le Nouvel An chinois.*

PAGE 94, *image 205* ART DIRECTOR: *David Curtis* DESIGNER: *Joan Bittner* PHOTOGRAPHER: *Steve Underwood* ARTIST/ILLUSTRATOR: *David Curtis* AGENCY: *Curtis Design* CLIENT: *Ghirardelli Chocolate Company* COUNTRY: *USA* PRINTER: *Lawson Mardon Label* TYPEFACE: *Casteller, M Bembo, Ribbon* PAPER: *65 lb. non-metallized foil* ■ *This package, created for a manufacturer and marketer of chocolates, con-*

veys the image of a high quality gift item. ● *Geschenkverpackung für Schokolade bester Qualität.* ▲ *Emballage-cadeau pour un assortiment de chocolats surfins.*

PAGE 94, *image 206* ART DIRECTOR/DESIGNER: *Kate Fujishima* ARTIST/ILLUSTRATOR: *Linda Frichtel* AGENCY: *Murrie Lienhart Rysner and Associates* CLIENT: *Dayton-Hudson* COUNTRY: *USA* ■ *This decorative holiday tin was designed for gourmet candies.* ● *Eine dekorative Dose für Konfekt erster Qualität.* ▲ *Cette boîte décorative de Noël a été conçue pour des bonbons surfins.*

PAGE 94, *image 207* ART DIRECTOR: *Kenzo Nakagawa* DESIGNERS: *Kenzo Nakagawa, Nobuyane Hiroyasu, Norikami Satoshi* AGENCY: *NDC Graphics* CLIENT: *Tower Shop* COUNTRY: *Japan* PRINTER: *Tokyo Shiko Co., Ltd.* TYPEFACE: *Rotis Semi Sans* PAPER: *Coated 110g.* ■ *The package was designed with traditional typography that reads "Cho-ko-rei-to," a Japanese phoneticization of the English word "chocolate."* ● *Das englische Wort 'chocolate', in der phonetischen Schreibweise für Japaner: «Cho-ko-rei-to».* ▲ *L'emballage présente une typographie traditionnelle. "Cho-ko-rei-to" est une phonétisation du mot anglais «chocolat».*

PAGE 94, *image 208* ART DIRECTOR: *Annette Harcus* DESIGNERS: *Lucy Walker, Anette Harcus* PRODUCT PHOTOGRAPHER: *Keith Arnold* ARTIST: *Simon Penton and Drawing Book* AGENCY: *Harcus Design* CLIENT: *Cadbury Schweppes* COUNTRY: *New Zealand* PRINTER: *Cadbury Schweppes* TYPEFACE: *Hand-drawn, Copperplate Gothic, Engravers Bold* ■ *The packaging for chocolate coffee creams was designed to express an Italian heritage and the conviviality of café life.* ● *Hier ging es um den italienischen Ursprung des Cappuccino und das gesellige Beisammensein in Cafés.* ▲ *Il s'agissait de faire ressortir la convivialité inhérente à la vie des cafés et les origines italiennes du cappuccino.*

PAGE 94, *image 209* ART DIRECTOR/PHOTOGRAPHER: *Jean-Jacques Schaffner* DESIGNER/ARTIST/ILLUSTRATOR: *Silvana Conzelmann* AGENCY: *Schaffner & Conzelmann AG* CLIENT: *Coop Schweiz* COUNTRY: *Switzerland* PRINTER: *GBC/Birkhäuser* ■ *Package design for an extension of the Havelaar product range. The Max Havelaar foundation is committed to paying fair prices to the suppliers of raw materials.* ● *Packungsgestaltung für eine Sortimentserweiterung der Havelaar-Produkte. Die Max-Havelaar-Stiftung tritt für eine faire Bezahlung von Rohstoffen ein.* ▲ *Design d'emballage pour une extension de la gamme des produits Havelaar. La fondation Max Havelaar est réputée pour payer des prix honnêtes aux pays producteurs de fèves de cacao.*

PAGE 95, *image 210* ART DIRECTOR: *Alan Chan* DESIGNERS: *Alan Chan, Peter Lo* AGENCY: *Alan Chan Design Company* CLIENT: *Toblerone* COUNTRY: *Hong Kong* ■ *This packaging features the "triangle" look of the product and was created to be a special festival gift pack.* ● *Die typische Dreiecksform der Toblerone, hier als Geschenkverpackung für Hongkong.* ▲ *L'emballage triangulaire typique du Toblerone. Ici, comme emballage-cadeau destiné au marché de Hong Kong.*

PAGE 96, *image 211* ART DIRECTOR: *Alan Chan* DESIGNERS: *Alan Chan, Peter Lo* AGENCY: *Alan Chan Design Company* CLIENT: *Mr. Chan Tea Room* COUNTRY: *Hong Kong* ■ *This packaging is intended to express passion for a re-discovery of art and culture in contemporary life.* ● *Bei dieser Verpackung ging es um die Wiederentdeckung von Kunst und Kultur in der heutigen Zeit.* ▲ *Cet emballage exprime la passion pour la redécouverte de l'art et de la culture à notre époque.*

PAGE 96, *image 212* ART DIRECTOR: *Alan Chan* DESIGNERS: *Alan Chan, Peter Lo* AGENCY: *Alan Chan Design Company* CLIENT: *Mr. Chan Tea Room* COUNTRY: *Hong Kong* ■ *The inspiration for this range of packaging came from historical Asian imagery. The crackle finish on the canisters was used to evoke a contemporary, nostalgic mood.* ● *Basis dieser Packungsfamilie sind historische asiatische Bilder, wobei das Krakelee auf den Weissblechdosen die nostalgische Wirkung unterstützen soll.* ▲ *Cette famille d'emballages s'inspire d'images asiatiques historiques. La finition craquelée sur les boîtes en fer blanc confère une atmosphère empreinte de nostalgie.*

PAGE 97, *images 213–217* ART DIRECTOR/DESIGNER: *Steve Sandstrom* AGENCY: *Sandstrom Design* CLIENT: *Tazo Tea Company* COUNTRY: *USA* TYPEFACE: *Nickolas Cochin, Nuptial, Garamond No 3* ■ *This packaging was designed for a producer of select hot and loose teas, tea and fruit juice bottled beverages and concentrates.* ● *Packungsgestaltung für einen Händler ausgewählter Teesorten, der auch Tee- und Fruchtgetränke in Flaschen sowie Konzentrate anbietet.* ▲ *Cet emballage a été conçu pour un producteur de thés de premier choix qui propose également des thés et des jus de fruits en bouteille et des concentrés.*

PAGE 98, 99; *images 218, 219* ART DIRECTOR: *Tassilo von Grolman* DESIGNER: *Birgit Osterhage* PRODUCT PHOTOGRAPHER: *Norbert Latocha* ARTIST/ILLUSTRATOR: *Alan Fletcher* AGENCY: *Tassilo von Grolman Design* CLIENT: *Club English Tea* COUNTRY: *Germany* ■ *In view of their 300th anniversary in the year 2006, Twinings has initiated the revival of the tea caddy, a porcelain bowl used by the upper class in Great Britain before tea was sold in tin cans. The new collection of tea caddies will be produced by Rosenthal in a limited edition of 2006. While Tassilo von Grolman designed the tea caddy itself, each year well-known artists will put their personal imprint on the various editions.* ● *Im Hinblick auf ihr dreihundertjähriges Bestehen im Jahre 2006 liess das britische Teehandelshaus Twinings den traditionellen 'tea caddy' wieder aufleben, ein Porzellangefäss, das vor allem der britis-*

chen Oberschicht zur Aufbewahrung ihres Tees diente, bevor dieser in Blechdosen verkauft wurde. Die neue Kollektion von tea caddies, wird exklusiv von Rosenthal in einer limitierten Auflage von insgesamt 2006 Stück produziert. Die Form wurde von Tassilo von Grolman entworfen, während bekannte Künstler alljährlich eine Edition durch ihre persönliche Handschrift prägen werden. ▲ *Dans l'optique de son 300ème anniversaire en 2006, la maison de thé Twinings a fait revivre le 'tea caddy', un récipient en porcelaine, utilisé autrefois par l'aristocratie britannique pour conserver du thé avant que celui-ci ne soit vendu dans des boîtes en fer-blanc. Rosentahl produira en exclusivité la nouvelle collection de tea caddies qui comptera en tout 2006 pièces. Le designer allemand Tassilo von Grolman a conçu la forme du tea caddy, tandis que chaque année, de nouveaux designers y apporteront leur touche personnelle.*

PAGE 99, *image 220* ART DIRECTOR: *Tassilo von Grolman* DESIGNER: *Thomas Bech* PRODUCT PHOTOGRAPHER: *Norbert Latocha* ILLUSTRATOR: *Thomas Bech* AGENCY: *Tassilo von Grolman Design* CLIENT: *Club English Tea* COUNTRY: *Germany* PRINTER: *Hellweg Kartonagen* ■ *The fresh blue waterdrops on these ice tea glasses suggest sparkling freshness. The carton containing a set of six glasses and also picks up this design.* ● *Die frischen, blauen Wassertropfen auf diesen Twinings-Eisteegläsern unterstreichen den prickelnde Kühle, die Eistee-Liebhaber von diesem Getränk erwarten. Auf dem sechs Gläser enthaltenden Karton wiederholt sich das Dekor der Teegläser.* ▲ *Les gouttes d'eau bleues sur ses verres à thé glacé Twinings soulignent l'agréable fraîcheur que procure la consommation de cette boisson. Le carton contenant six verres reprend les motifs des verres.*

PAGE 99, *image 221* CREATIVE DIRECTOR/DESIGNER: *Amy Knapp* EXECUTIVE CREATIVE DIRECTOR: *Peter Allen* ILLUSTRATOR: *Heather Preston* CLIENT: *DFS Merchandising Ltd.* COUNTRY: *Taiwan* PAPER: *SBS Board with Matte Lamination* ■ *The design goal was to convey an image of traditional and gourmet Chinese food packaging.* ● *Die traditionelle Verpackung für chinesische Gourmet-Produkte war die Basis für diesen Entwurf.* ▲ *Emballage traditionnel pour des produits chinois destinés aux gourmets.*

PAGE 99, *image 222* ART DIRECTOR: *Nigel Bullivant* DESIGNER: *Helen Howat* ARTIST/ILLUSTRATOR: *Gordon Hurden* AGENCY: *Vineyard Design Limited* CLIENT: *J Sainsbury plc.* COUNTRY: *United Kingdom* PRINTER: *Lawson Mardon Packaging* TYPEFACE: *Times Bold* ■ *The aim was to create a distinctive, collectable, contemporary design with gift appeal.* ● *Hier ging es um den Entwurf einer speziellen, modernen Verpackung, die man gerne aufhebt oder verschenkt.* ▲ *Le but était de créer un emballage spécial au design contemporain que l'on a envie de collectionner ou d'offrir en cadeau.*

PAGE 99, *image 223* ART DIRECTOR/DESIGNER: *Yasuo Tanaka* AGENCY: *Package Land Co, Ltd.* CLIENT: *Tamaya Co. Ltd.*

PAGE 99, *image 224* CREATIVE DIRECTOR: *Kan Tai-keung* ART DIRECTOR: *Freeman Lau Siu Hong* DESIGNERS: *Freeman Lau Siu Hong, Eddy Yu Chi Kong* AGENCY: *Kan Tain-keung Design & Associates* CLIENT: *Unilever Hong Kong Ltd.* COUNTRY: *Singapore* ■ *The new packaging aims to revitalize an old image and to prepare for the product's entry into the Chinese market.* ● *Aufgabe dieser neuen Verpackung war die Belebung des Images und die Lancierung des Produktes auf dem chinesischen Markt.* ▲ *Objectif de ce nouvel emballage: lifting de l'image du produit et lancement sur le marché chinois.*

PAGE 100, *image 225* ART DIRECTOR: *Barrie Tucker* DESIGNERS: *Jody Tucker, Claire Rose* PRODUCT PHOTOGRAPHER: *Simon Vaughn* AGENCY: *Tucker Design* CLIENT: *Lactos* COUNTRIES: *Australia (small packs), Holland (large pack wrapper)* PRINTER: *Mercury Walch, Tasmania, Australia* HERITAGE LOGO: *Hand artwork; (type): Caslon Open Face, Bodoni* ■ *This packaging was designed for a range of Tasmanian cheeses.* ● *Packungsgestaltung für ein Käsesortiment aus Tasmanien.* ▲ *Emballage conçu pour un assortiment de fromages tasmaniens.*

PAGE 101, *image 226* ART DIRECTOR: *David Ceradini* DESIGNER: *Natalie Jacobs* ARTIST/ILLUSTRATOR: *Tim Shannon* AGENCY: *Ceradini Design* CLIENT: *The Great Atlantic & Pacific Tea Co.* HAND LETTERING: *Hon Leong*

PAGE 102, *images 227, 228* ART DIRECTOR/DESIGNER: *Peter Schmid* PRODUCT PHOTOGRAPHER: *Robert Striegl* AGENCY: *Fölser + Shernhuber* CLIENT: *Anton Bürstunger Landfleischerei* COUNTRY: *Austria* PRINTER: *Esterman* TYPEFACE: *Kipp, Kabel, LeChochin* PAPER: *Serilux 2000* ■ *The pun underlying the product name, "Landstreicher," is based on the German word for a person who roams around the countryside. The "Streicher" part of the word also refers to a "spread"—as in the meat spread.* ● *Der Markenname «Landstreicher» dient hier als Wortspiel im Zusammenhang mit dem Produkt, einer Leberstreichwurst.* ▲ *Marque d'un pâté de foie autrichien. L'allemand joue sur le mot «Landstreicher» qui signifie tout à la fois «vagabond» et «produit à tartiner».*

PAGE 103, *images 229, 230* ART DIRECTOR/DESIGNER: *Hansjörg Bolt* ARTIST/ILLUSTRATOR: *Heike Grein* AGENCY: *Bolt, Koch & Co.* CLIENT: *Volg Konsumwaren AG* COUNTRY: *Switzerland* ● *Brand development and package design for the in-house labels of a food chain.* ● *Markenentwicklung und Verpackungsgestaltung für die Eigenmarken einer Schweizer Lebensmittelkette.* ▲ *Développement de marque et design du packaging des produits d'une chaîne de magasins d'alimentation suisse.*

PAGE 104, *image 231* ART DIRECTOR/DESIGNER: *Danny Klein* PHOTOGRAPHY: *Studio voor Fotografie* ARTIST/ILLUSTRATOR: *Raymond Lobato* AGENCY: *Millford-Van den Berg Design* CLIENT: *Albert Heijn B.V.* ■ *This design utilizes the client's standard graphic design codes in its choice of colors and illustration.* ● *Die Auswahl der Farben und Illustrationen dieser Verpackung steht im Einklang mit den typischen graphischen Elementen des Auftraggebers.* ▲ *Le choix des couleurs et des illustrations de cet emballage reprennent les éléments graphiques traditionnels de Cartier.*

PAGE 105, *image 232* ART DIRECTOR: *Dennis Thompson* DESIGNER: *Jody Thompson* ARTIST/ILLUSTRATOR: *Leland Klanderman* AGENCY: *Thompson Design Group* CLIENT: *Pacific Grain Products* COUNTRY: *USA* PAPER: *Recycled cardboard* ■ *Designed for a client that produces a variety of natural grain products, this packaging is intended to convey qualitites of natural goodness and health.* ● *Das Thema dieser Verpackung sind die natürlichen, gesunden Getreideprodukte des Auftraggebers.* ▲ *Des ingrédients sains et naturels, tel est le message de ce packaging conçu pour une ligne de produits diététiques.*

PAGE 105, *image 233* ART DIRECTOR/DESIGNER: *Primo Angeli, Richard Sheve* AGENCY: *Primo Angeli* CLIENT: *Arrowhead Mills* COUNTRY: *USA* ■ *The objectives were to establish a platform for branding, to communicate the product descriptor clearly, and to command high self-impact. Originally available only in health food stores, the new package is intended to allow the product to compete in an open supermarket environment.* ● *Das ursprünglich nur in Reformhäusern erhältliche Produkt sollte auch im Supermarktgestell bestehen können. Aufgabe der Verpackung war daher ein markanter, eigenständiger Auftritt.* ▲ *Nouvel emballage visant à assurer le succès en grandes surfaces d'un produit commercialisé à l'origine dans les magasins de produits diététiques uniquement.*

PAGE 105, *image 234* ART DIRECTOR/DESIGNER: *Bill Chiaravalle* AGENCY: *Landor Associates* CLIENT: *J.M. Smuckers* COUNTRY: *USA*

PAGE 105, *image 235* ART DIRECTOR: *Jon Weden* DESIGNER: *Martha Furman* ARTIST/ILLUSTRATOR: *Barbara Kelly* AGENCY: *Landor Associates* CLIENT: *H.E.B.* ■ *The task was to create a line of packaging which fits the client's branding strategy while conveying wholesome, flavorful qualites.* ● *Entwicklung einer Verpackungslinie, die den fruchtigen Geschmack der Konfitüre zum Ausdruck bringen und gleichzeitig die Markenstrategie des Kunden berücksichtigen sollte.* ▲ *Conditionnement des confitures H.E.B. Il s'agissait ici de respecter la stratégie de marque du client et d'évoquer la saveur fruitée du produit.*

PAGE 105, *image 236* ART DIRECTOR/DESIGNER: *Angelo Sganzerla* ARTIST/ILLUSTRATOR: *Alfonso Goi* AGENCY: *Angelo Sganzerla* CLIENT: *Andrea Stainer* ■ *This packaging promotes a specialty food item made from Piemontese hazelnuts.* ■ *Verpackung für eine Haselnuss-Spezialität aus dem Piemont.* ▲ *Emballage d'une spécialité piémontaise à base de noisettes.*

PAGE 105, *image 237* ART DIRECTOR: *David Ceradini* DESIGNER: *Lori Raymer* PRODUCT PHOTOGRAPHER: *John Uber* ARTIST/ILLUSTRATOR: *Tim Shannon* AGENCY: *Ceradini Design* CLIENT: *The Great Atlantic & Pacific Tea Co.* ■ *The assignment was to create a hearty, distinctive image for a 100% pure beef product.* ● *Hier ging es um ein herzhaftes Image und den eigenständigen Auftritt eines Produktes aus 100% Rindfleisch.* ▲ *L'objectif était de créer une image forte qui permette à ce produit 100% pur bœuf de se démarquer de la concurrence.*

PAGE 105, *image 238* ART DIRECTOR: *Primo Angeli, Brody Hartman* TYPOGRAPHERS: *Brody Hartman, Sherry Brigham* COMPUTER ILLUSTRATORS: *Devin Muldoon, Liz Wheaton* AGENCY: *Primo Angeli* CLIENT: *Del Monte* COUNTRY: *USA* ■ *The packaging relies on attention to details in typography and illustration to denote tradition and fresh flavor. The product's Italian authenticity is emphasized through the high-recognition Del Monte name.* ● *Liebe zum Detail in der Beschriftung und Illustration dienten dazu, Tradition und Frische zum Ausdruck zu bringen. Die Bekanntheit des Markennamens Del Monte reichte, um den italienischen Charakter des Produktes hervorzuheben.* ▲ *Tradition et fraîcheur étaient les valeurs qu'il convenait de faire ressortir, ici au moyen d'une typographie soignée et des sujets choisis. Le nom de la marque Del Monte suffisait à lui seul à attester de l'origine italienne du produit.*

PAGE 105, *image 239* ART DIRECTORS: *Debbie Smith, Steven Addis* DESIGNER: *Debbie Smith* ARTIST/ILLUSTRATOR: *Robert Evans* AGENCY: *Monnens-Addis Design* CLIENT: *Bell-Carter Foods* COUNTRY: *USA* ■ *In order to position the product as a premium brand a design reminiscent of old fruit crate graphics conveys a sense of heritage while infusing it with a fresh—rather than canned—image.* ● *Durch den Einsatz der an alte Obstkisten erinnernden Graphik wird der Eindruck von Tradition und einem hochwertigen Produkt erzeugt, wobei der Verbraucher eher an ein frisches als an ein konserviertes Erzeugnis denken soll.* ▲ *Le design évoque les notions de tradition et de qualité et tend à présenter le produit comme un produit frais plutôt que comme une conserve.*

PAGE 106, *image 240* ART DIRECTOR/DESIGNER: *G.J. Babr* PRODUCT PHOTOGRAPHER: *Studio CP Hoffmann* AGENCY: *Package & Logo Design* CLIENT: *Appel & Frenzel GmbH* ■ *Package design for the launch of a range of pâté de foie gras on the German market.* ● *Verpackung für die Lancierung eines hochpreisigen*

Gänselebersortiments. ▲ *Packaging créé pour le lancement de pâtés de foie gras de qualité supérieure sur le marché allemand.*

PAGE 107, *image 241* ART DIRECTOR/DESIGNER/ILLUSTRATOR: *Sibylle Haase* AGENCY: *Atelier Haase & Knels* CLIENT: *B. Grashoff Nachf.* COUNTRY: *Germany* PRINTER: *Nord-Etikett GmbH* ● *Label design for high quality jams.* ■ *Gestaltung von Etiketten für Feinkost-Marmeladen.* ▲ *Conception d'étiquettes pour des confitures de qualité supérieure.*

PAGE 108, 109; *images 242–245* ART DIRECTOR: *Kit Hinrichs* DESIGNER: *Jackie Foshaug* AGENCY: *Pentagram Design* CLIENT: *Columbus Salame* ■ *The main objective was to update and revitalize the client's logo using the primary color of their previous logo. The wood-cut style of the illustration was employed to achieve a classic look/* ● *Ziel dieser Verpackungsgestaltung war eine Auffrischung des Firmenlogos, wobei die Grundfarbe des vorherigen Logos übernommen wurde. Als Illustration wurde ein Holzschnitt verwendet, um einen klassischen Look zu erzielen.* ▲ *Cet emballage devait rajeunir le logo d'entreprise. La couleur de base de l'ancien logo a été conservée. En guise d'illustration, une gravure sur bois afin de conférer une touche de classicisme au produit.*

PAGE 110, *image 246* ART DIRECTOR/DESIGNER: *Tom Antista, Thomas Fairclough* PRODUCT PHOTOGRAPHY: *Michael West Photography* AGENCY: *Antista Fairclough* CLIENT: *Wolfgang Puck Food Co.* COUNTRY: *USA* ■ *A proprietary label for Spago, this design was developed for use at the restaurant in California.* ● *Flaschengestaltung für Spago-Olivenöl, das für ein Restaurant in Kalifornien bestimmt ist.* ▲ *Bouteille conçue pour l'huile d'olive Spago, destinée à un restaurant californien.*

PAGE 110, *image 247* ART DIRECTOR/DESIGNER: *Patti Britton* PRODUCT PHOTOGRAPHER: *W.J. Wickham* AGENCY: *Britton Design* CLIENT: *Viansa Winery* COUNTRY: *USA* PRINTER: *Custom Label.* TYPEFACE: *Augusta Inline, Centaur* ■ *To reflect the already existing style of the client's renaissance-style wine labels, the new food labels incorporate the two-label look in order to reveal the product for the consumer. A 15th century fresco from the winery walls is used for the die-cut upper label.* ● *Bei diesem neuen Produkt des Auftraggebers wurde wie bei seinen bereits gut eingeführten Weinen mit Etiketten im Stil der Renaissance ebenfalls mit zwei Etiketten gearbeitet. Im oberen Etikett ist ein Fresko aus dem 15. Jahrhundert zu sehen, das sich an der Wand der Weinkellerei befindet.* ▲ *Conçues pour un nouveau produit, ces étiquettes rappellent les étiquettes de style Renaissance créées avec succès pour les vins du producteur. L'étiquette supérieure illustre une fresque du XVe siècle qui orne l'un des murs des caves.*

PAGE 111, *images 248, 249* ART DIRECTOR: *Marion English, Terry Slaughter* DESIGNERS: *Marion English, Rebecca Fulmer* PRODUCT PHOTOGRAPHER: *Bruce Sutherland* AGENCY: *Slaughter-Hanson* CLIENT: *Bottega Restaurant/Cafe* COUNTRY: *USA* PRINTER: *Ad Shop* TYPEFACE: *Venetian* PAPER: *dull/white mac-tac* ■ *These bottles were designed to be displayed and sold out of an upscale Italian restaurant/cafe; the task was to create an unusual, classic design that also reflected the ambiance and humor of the restaurant environment.* ● *Diese Flaschen sind für den Verkauf in italienischen Restaurants/Cafés der gebobenen Preisklasse bestimmt. Es galt, ein ungewöhnliches, klassisches Design zu finden, das die Ambience des Restaurants reflektiert.* ▲ *Bouteilles conçues pour l'exposition et la vente dans un café-restaurant italien réputé; la tâche consistait à créer un design classique inhabituel qui reflète également l'ambiance des restaurants.*

PAGE 111, *image 250* ART DIRECTOR/DESIGNER: *Peter Schmid* PRODUCT PHOTOGRAPHER: *Robert Striegl* ARTIST/ILLUSTRATOR: *Mike Hirschl* AGENCY: *Fölser + Schernhuber* CLIENT: *Sonnentor Kräuterbandelsges. GmbH* COUNTRY: *Austria* PRINTER: *Esterman* TYPEFACE: *Old Claude, Caslon Antique* PAPER: *JAC Script* ■ *This packaging and identity design were created for a complete range of organically grown products packed by hand by farmers' families.* ■ *Neugestaltung für eine umfangreiche Serie von biologisch angebauten Erzeugnissen, die auf den Höfen selbst von den Bauernfamilien verpackt werden.* ▲ *Packaging de toute une gamme de produits diététiques issus de cultures biologiques. Le conditionnement est fait à la main sur le lieu de production.*

PAGE 111, *image 251* ART DIRECTOR/DESIGNER/ARTIST/ILLUSTRATOR: *Astrid Becher* PRODUCT PHOTOGRAPHER: *Peter Hernandez* AGENCY: *Jacobson Rost* CLIENT: *Schreier Malting Co.* COUNTRY: *USA* PRINTER: *Bemis Company, Inc.* TYPEFACE: *Bodega* PAPER: *Kraft Paper* ■ *These storage and shipping bags were created to hold 50 lb. of Malt.* ● *Lager- und Verschiffungssäcke für Malz mit einem Fassungsvermögen von ca. 20 kg.* ▲ *Sacs d'une contenance de 20 kg environ, destinés au conditionnement et à l'expédition du malt.*

PAGE 112, *images 252–254* CREATIVE DIRECTOR: *Amy Knapp* EXECUTIVE CREATIVE DIRECTOR: *Peter Allen* DESIGNER: *Amy Knapp, Jane Campbell* PRODUCT PHOTOGRAPHER: *Deborah Jones* ILLUSTRATOR: *Justin Carroll* CLIENT: *DFS Merchandising Ltd.* COUNTRIES: *Australia, US* PRINTERS: *Various* ■ *The design needed to encompass all categories of food products for the Australia region, so the Koala Logo was developed to change according to category. The look features traditional Australian characteristics with a whimsical Koala as its focus.* ● *Verpackungssystem für ein ganzes Sortiment von Produkten, die für den australischen Markt bestimmt sind. Die Varianten des Koala-Logos kennzeichnen die verschiedenen Produktkategorien.* ▲ *Ligne d'emballages créée pour une vaste* gamme de produits destinés au marché australien. Le logo, un koala décliné en plusieurs versions, indique la catégorie du produit.

PAGE 114, *image 255* ART DIRECTOR: *Lowell Williams* DESIGNERS: *Lowell Williams, Bill Carson, Matt Heck* ARTISTS/ILLUSTRATORS: *Jeff Williams, John Bohls* AGENCY: *Pentagram Design* CLIENT: *Doubletree Hotels Corporation* COUNTRY: *USA* PRINTER: *Olive Can Company* TYPEFACE: *Trajan* ■ This packaging was created for cookies available at the client's hotel properties. ● Packungsgestaltung für die Gebäck-Hausmarke einer Hotelkette. ▲ Emballage créé pour les biscuits d'une chaîne hôtelière.

PAGE 115, *image 256* ART DIRECTOR: *Jack Anderson* DESIGNERS: *Jack Anderson, John Anicker* PRODUCT PHOTOGRAPHER: *Tom Collicott* ILLUSTRATOR: *John Fretz* AGENCY: *Hornall-Anderson Design Works, Inc.* CLIENT: *Continental Mills* COUNTRY: *USA* PRINTER: *Bemis, Vancouver* TYPEFACE: *Matrix* ■ This packaging was created for a line of bread mix for use with the client's bread-making machine. ● Diese Verpackung enthält eine Brotbackmischung, die man in der Brotbackmaschine von Continental Mills verwenden soll. ▲ Cet emballage contient un mélange de pâtes à pain qu'il s'agit de préparer dans l'appareil ménager proposé par le client.

PAGE 116, *image 257* ART DIRECTOR: *Keith Steimel* DESIGNER/ ARTIST/ILLUSTRATOR: *Mike Endy* AGENCY: *Cornerstone* CLIENT: *Chebby Foods* COUNTRY: *USA* ■ The challenge was to develop a line of packaging which promotes the authenticity of the Southwestern salsa product, yet which also conveys a little more "attitude". ● Diese Verpackungslinie für eine Gewürzsaucen-Marke sollte einerseits die Herkunft aus dem Südwesten der USA betonen, andererseits dem Produkt eine eigene Persönlichkeit geben. ▲ Le packaging évoque l'authenticité et la qualité du produit – une sauce typique du Sud-Ouest américain.

PAGE 116, *image 258* ART DIRECTORS/DESIGNERS: *Tom Antista, Thomas Fairclough* PRODUCT PHOTOGRAPHER: *Michael West Photography* AGENCY: *Antista Fairclough Design* CLIENT: *K.T.'s Kitchen* COUNTRY: *USA* ■ In this design, the chef icon serves as an identifier of this gourmet pizza. ● Packungsgestaltung für eine Gourmet-Pizza, deren Markenzeichen der Koch ist. ▲ Packaging d'une marque de pizza américaine. Le logo – un chef de cuisine – est symbole de qualité.

PAGE 116, *image 259* ART DIRECTOR/DESIGNER: *Angelo Sganzerla* ARTIST/ ILLUSTRATOR: *Alfonso Goi* AGENCY: *Angelo Sganzerla* CLIENT: *Andrea Stainer*

PAGE 116, *image 260* ART DIRECTOR: *John Hornall* DESIGNERS: *John Hornall, Jana Nishi, Debra Hampton, Leo Raymundo, Mary Chin Hutchinson* PRODUCT PHOTOGRAPHER: *Tom McMackin* ARTIST/ILLUSTRATOR: *Carolyn Vibbert* AGENCY: *Hornall Anderson Design Works, Inc.* CLIENT: *Darigold* COUNTRY: *USA* PRINTER: *James River* TYPEFACE: *Futura* ■ This design was created to update the current butter packaging for the client's revised product line. ● Neugestaltung der Butterverpackung im Rahmen des überarbeiteten Auftritts des gesamten Produktsortiments des Auftraggebers. ▲ Nouvel emballage d'un beurre américain, créé dans le cadre d'un «lifting» complet des produits de la marque.

PAGE 116, *image 261* ART DIRECTOR: *Mark Oliver* DESIGNERS: *Patty Devlin-Driskel, Mark Oliver* PRODUCT PHOTOGRAPHER: *Bill Boyd* ARTIST/ILLUSTRATOR: *Carla Siboldi* AGENCY: *Mark Oliver, Inc.* CLIENT: *San Luis Sourdough* COUNTRY: *USA* PRINTER: *Huntsman Packaging* CUSTOM LETTERING: *Holly Dickens* ■ The new packaging was designed to reflect the line's premium quality and higher price, and to appeal to the upscale consumer. ● Diese neue Verpackung für ein typisch amerikanisches Gebäck sollte die hervorragende Qualität des Produktes zum Ausdruck bringen und den anspruchsvollen Konsumenten ansprechen. ▲ Nouvel emballage évoquant la qualité supérieure de ce produit qui vise des consommateurs exigeants au pouvoir d'achat élevé.

PAGE 116, *image 262* ART DIRECTOR/DESIGNER: *Maxey Andress* ARTISTS/ ILLUSTRATORS: *Clem Bedwell, Jay Montgomery, Lamar Smith* AGENCY: *EM2 Design* CLIENT: *Seckinger-Lee Company* COUNTRY: *USA* PRINTER: *J.L Clark, Waldorf* TYPEFACE: *Futura Book, Times Roman* ■ The purpose was to increase shelf-appeal and stocking committments with high-end retail purveyors. ● Diese Verpackung sollte durch einen wirkungsvolleren Auftritt des Produktes und bessere Lagerfähigkeit den Einzelhandel der gehobenen Preisklasse ansprechen. ▲ Conçu pour le commerce de luxe, ce packaging visait à renforcer l'attrait du produit à l'étalage et à en faciliter le stockage.

PAGE 116, *image 263* CREATIVE DIRECTOR/DESIGNER: *Amy Knapp* EXECUTIVE CREATIVE DIRECTOR: *Peter Allen* ILLUSTRATOR: *Heather Preston* CLIENT: *DFS Merchandising Ltd.* COUNTRY: *Taiwan* PAPER: *SBS Board with Matte Lamination* ■ The design needed to convey a look of traditional and gourmet Chinese food packaging for the products in the client's Taiwanese market. ● Bei der Verpackung dieser für den taiwanesischen Inlandmarkt bestimmten Produkte ging es um das Image von traditionellen chinesischen Gourmet-Produkten. ▲ Produits destinés au marché taïwanais. Le packaging suggère la finesse de ces spécialités chinoises.

PAGE 116, *image 264* ART DIRECTOR/DESIGNER: *Bruce Yelaska* PRODUCT PHOTOGRAPHER: *Tom McCarthy* AGENCY: *Bruce Yelaska Design* CLIENT: *Nonni's Biscotti* COUNTRY: *USA* PRINTER: *Everett Graphics* TYPEFACE: *Hand lettering, Clearface Gothic, Sabon* ■ This package for a biscotti bakery was designed to work within a system of packaging for gourmet biscotti. ● Packungsgestaltung inner-

halb eines Packungssystems für die Spezialität einer Bäckerei. ▲ Packaging pour les spécialités d'une boulangerie-pâtisserie industrielle.

PAGE 117, *image 265* ART DIRECTOR: *Helmut Rottke* DESIGNER/ARTIST/ILLUSTRATOR: *Ulrike Jägerfeld* AGENCY: *Rottke Werbung* CLIENT: *Irish Dairy Board* COUNTRY: *Ireland* PRINTER: *Drukkerij Tulp Zwolle Zwolle, Netherlands* TYPEFACE: *Celtic Letters* PAPER: *Tissue paper 12-14g.* ● New package design for the launch of a cheese from Ireland. ■ Neugestaltung der Verpackung für die Einführung einer Käsespezialität aus Irland. ▲ Emballage créé pour le lancement d'un fromage irlandais.

PAGE 118, *images 266–269* ART DIRECTOR: *Supon Phornirulit* DESIGNER/ARTIST/ ILLUSTRATOR: *Apisak Saibua* PRODUCT PHOTOGRAPHER: *Oi Jakrarat Veerasarn* AGENCY: *Supon Design Group Inc* CLIENT: *Fresh Market* COUNTRY: *Typeface: Arrow* PAPER: *Crack-n-Peel* ■ This label system was created for supermarket produce. ● Etikettgestaltung für die Produkte eines Supermarktes in Hongkong. ▲ Etiquettes créées pour les produits d'un supermarché de Hongkong.

PAGE 119, *image 270* ART DIRECTOR/DESIGNER: *Angelo Sganzerla* ARTIST/ ILLUSTRATOR: *Alfonso Goi* AGENCY: *Angelo Sganzerla* CLIENT: *L'Erbaolario* ■ This packaging was created to emphasize the qualities of the client's naturally derived products. ● Bei der Verpackung ging es um die Betonung der natürlich Ingredienzen der Shampoos, Parfums und Körperpflegemittel. ▲ L'emballage met l'accent sur les qualités des produits fabriqués naturellement.

PAGE 119, *image 271* ART DIRECTOR: *Georges Lachaise* DESIGNER: *Henri Champy* PRODUCT PHOTOGRAPHY: *Studio Appi, Klaus Ramshorn* AGENCY: *George Lachaise Design* CLIENT: *Elie-Arnaud Denoix* COUNTRY: *France* PRINTER: *Georges Lachaise* ■ This package was designed for the launch of a new product. ● Packungsgestaltung für die Einführung neuer Produkte. ▲ Emballage conçu pour le lancement de nouveaux produits.

PAGE 119, *image 272* ART DIRECTOR: *Per Magne Lund* DESIGNER: *Runa Fridén* ARTIST/ILLUSTRATOR: *Christopher Wormel* AGENCY: *Christensen/Lund* CLIENT: *Regal Mølle, Starbburet* COUNTRY: *Norway* PRINTER: *Lyche, Mandal Pose* TYPEFACE: *Kabel* ■ This packaging was created for a series of different flour types. ● Verpackung für vier verschiedene Mehlsorten. ▲ Emballage créé pour une gamme comprenant plusieurs sortes de farine.

PAGE 119, *image 273* ART DIRECTOR/DESIGNER: *Per Magne Lund* ARTIST/ ILLUSTRATOR: *Kjell Nupen* AGENCY: *Christensen/Lund* CLIENT: *Norske Meierier (Tine)* COUNTRY: *Norway* PRINTER: *Tronheim Eske, Systemetikettering* TYPEFACE: *Gill* ■ This packaging series for cheeses with "family branding" features oil paintings by the famous Norwegian artist, Kjell Nupen. ● Packungsfamilie für eine Reihe von Käsesorten, mit Reproduktionen von Ölgemälden des norwegischen Malers Kjell Nupen. ▲ Série d'emballages à fromage ornés de reproductions de toiles du peintre norvégien Kjell Nupen.

PAGE 119, *image 274* ART DIRECTOR/DESIGNER: *Ed Johnson* PHOTOGRAPHY: *Jeff Michaels Photography* ARTIST/ILLUSTRATOR: *Shirley Chapman* AGENCY: *Gaylord Graphics* CLIENT: *California Tomato Pickers* COUNTRY: *USA* PRINTER: *Gaylord Graphics* TYPEFACE: *Variex Regular* PAPER: *42 lb. Mottled White, 100% Rod Coated White* ■ This client wanted to visually attract the Costco/Price Club consumer. The texture and movement of the design and illustration reflect the idea of a "fast-moving" consumer product. ● Mit dieser Verpackung sollten die Kunden eines Anbieters preiswerter Massenartikel angesprochen werden. Der gesamte Auftritt suggeriert ein Produkt mit schnellem Umsatz. ▲ Packaging accrocheur destiné à des articles de consommation de masse bon marché. Le concept global reflète l'idée d'un «produit de consommation à forte rotation».

PAGE 119, *image 275* ART DIRECTOR/DESIGNER: *Hansjörg Bolt* ARTIST/ILLUSTRATOR: *Heike Grein* AGENCY: *Bolt, Koch & Company.* CLIENT: *Volg Konsumwaren AG* COUNTRY: *Switzerland* ■ Brand identity and package design for the in-house brands of a food chain. ● Markenentwicklung und Packungsgestaltung für Eigenmarken einer Lebensmittelkette. ▲ Développement de l'identité de marque et du packaging pour les marques de distributeur d'une chaîne de magasins d'alimentation.

PAGE 119, *image 276* ART DIRECTOR: *Rita Damore* DESIGNERS: *Rita Damore, Doug Gilmour, Rob Mesarchik* PHOTOGRAPHER: *Deborah Jones* ARTIST/ILLUSTRATOR: *Rob Mesarchik* AGENCY: *Damore-Johann Design* CLIENT: *Philchick Inc.* COUNTRY: *USA* PRINTER: *Standard Paper Box* TYPEFACE: *Modula, Copperplate* PAPER: *SBS 18 pt.* ■ The packaging emphasizes the addition of herb flavoring to the crust to strengthen the product's market position. The photography enhances the appetite appeal of the product and its gourmet ingredients. The graphics reinforce these qualities while at the same time establishing a graphic system for future line extensions. ● Diese Verpackung, die sich auch für die zukünftige Erweiterung der Linie eignen muss, betont die Beigabe von Kräutern, um die Marktposition des Produktes zu stärken. Photographie und Graphik unterstützen den Anspruch eines appetitlichen Feinschmeckerproduktes. ▲ L'emballage devait également se prêter à un futur élargissement de la gamme. L'accent est mis sur l'ajout de fines herbes aux recettes pour renforcer la position du produit sur le marché. Photographie et graphisme mettent en valeur le côté appétissant de ces spécialités gourmandes.

PAGE 119, *image 277* ART DIRECTOR/DESIGNER: *Hansjörg Bolt* ARTIST/ILLUSTRATOR: *Heike Grein* AGENCY: *Bolt, Koch & Co.* CLIENT: *Volg Konsumwaren AG* COUNTRY: *Switzerland* ■ *Development of the brand identity and packaging for in-house brands of a food chain.* ● *Markenentwicklung und Verpackungsgestaltung für Eigenmarken einer Schweizer Lebensmittelkette.* ▲ *Développement de l'identité de marque et du packaging pour les marques de distributeur d'une chaîne de magasins d'alimentation.*

PAGE 120, 121; *images 278–281* ART DIRECTOR/ARTIST/ILLUSTRATOR: *Ian Lidji* DESIGNERS: *Alan Lidji, Janet Cowling* PRODUCT PHOTOGRAPHER: *Lynn Sugarman* PRODUCT PHOTOGRAPHER: *Jim Olvera* AGENCY: *Lidji Design Office* CLIENT: *Aromance Home Fragrances* COUNTRY: *USA* PRINTER: *Pelikan Press* TYPEFACE: *Minion* ■ *This high-end home fragrance packaging was designed for the low-end impulse buyer.* ● *Diese Verpackung für ein Raumparfum sollte den preisbewussten, spontanen Käufer ansprechen.* ▲ *Emballage d'un parfum d'intérieur destiné à favoriser les achats spontanés.*

PAGE 122, *image 282* ART DIRECTOR: *Glenn Tutssel* DESIGNER: *Garrick Hamm* AGENCY: *Tutssels* CLIENT: *Boots* COUNTRY: *Great Britain* TYPEFACE: *Sabon* ■ *For the packaging of these travel tissues, imagery depicting trains, boats, and planes was employed to evoke all the the fun and excitement of going on holiday.* ● *Das Thema dieser Verpackung für Reise-Erfrischungstücher sind Ferienfreuden, dargestellt durch Züge, Schiffe und Flugzeuge.* ▲ *Trains, bateaux et avions. L'excitation des voyages, tel est le thème illustré par les pochettes de ces serviettes rafraîchissantes.*

PAGE 123, *image 283* ART DIRECTOR: *Gertraud Hilbert* AGENCY: *Werbeatelier Fick Werbeagentur GmbH* CLIENT: *Rosenthal AG* COUNTRY: *Germany* ■ *The package design and the complete visual image of the glass series "diVino", a subbrand of Rosenthal targeted at the younger consumer, is based on the shape of the rhombus and its half form, the triangle. Shown are gift packages containing 2 each of three different types of glasses.* ● *Basis dieser Trinkglas-Serie «diVino», eine Rosenthal-Speziallinie für jüngere Zielgruppen, ist die Rautenform und ihre halbe Grundform, das Dreieck. Hier die Geschenkverpackungen mit je 2 Stück von drei verschiedenen Gläsern.* ▲ *Le concept graphique de l'emballage et l'image visuelle de la série de verres «diVino», une sous-marque de Rosenthal ciblant les jeunes consommateurs, se basent sur les formes d'un rhombe et sa demi-forme, le triangle. Ici, des pochettes-cadeau contenant chacune deux des trois types de verres fabriqués.*

PAGE 124, *image 284* ART DIRECTOR: *Tucker Viemeister* DESIGNERS: *Debbie Hahn, Stephanie Kim, Nick Graham* PRODUCT PHOTOGRAPHER: *Peter Medilek* AGENCY: *Smart Design* CLIENT: *Timex: Joe Boxer* COUNTRY: *USA* TYPEFACE: *Janson* ■ *The smiley face boxes, which protect the product and provide a souvenir, were designed to promote a line of watches.* ● *Diese Smiley-Boxen dienen als Souvenir und Werbung für eine Uhrenlinie, der sie gleichzeitig als Verpackung dienen.* ▲ *Conçues pour lancer une collection de montres, ces boîtes «smiley» font office d'«écrins» et peuvent être conservées en souvenir.*

PAGE 125, *image 285* ART DIRECTOR/DESIGNER: *Beth Parker* PRODUCT PHOTOGRAPHER: *Andrew Swaine* AGENCY: *Phillips Design Group* CLIENT: *Atlantic Technology* COUNTRY: *USA* PRINTER: *Rand Whitney* TYPEFACE: *Syntac* PAPER: *E-flut, Federal SBS* ■ *This line of packaging for computer accessories was designed to lighten up the products' image and to help them stand out in the chaotic environment of the typical computer superstore.* ● *Verpackungsgestaltung für Computer-Zubehör. Hier ging es darum, dem Image des Produktes mehr Leichtigkeit und eine unverwechselbare Präsenz im chaotischen Umfeld des typischen Computer-Ladens zu geben.* ▲ *Ligne d'emballages pour des accessoires pour ordinateur. L'objectif était d'améliorer l'image des produits et leur permettre de se démarquer dans l'environnement chaotique des magasins informatiques.*

PAGE 125, *image 286* ART DIRECTOR: *Michael Osborne* DESIGNER/ARTIST/ILLUSTRATOR: *Kristen Clark* PRODUCT PHOTOGRAPHER: *Tony Stromberg* AGENCY: *Michael Osborne Design* CLIENT: *Empire Berol USA* COUNTRY: *USA* PRINTER: *Watermark Press* TYPEFACE: *Label: Bureau Agency, Bank Gothic* ■ *The new identity and packaging were created to reflect Prismacolor's redesign of its double-ended art markers.* ● *Eine neue Markenidentität und Verpackung für einen verbesserten Künstlerstift von Prismacolor, der sich von zweiten Seiten benutzen lässt.* ▲ *Nouvelle identité de marque et packaging revisité pour les nouveaux stylos Prismacolor à double pointe.*

PAGE 126, *image 287* ART DIRECTOR/DESIGNER: *Louise Fili* PRODUCT PHOTOGRAPHER: *Ed Spiro* AGENCY: *Louise Fili Ltd.* CLIENT: *El Paso Chile Co.* COUNTRY: *USA* PRINTER: *Guynes Printing* TYPEFACE: *Nicolas Chochin* PAPER: *Simpson Filare* ■ *The goal was to package a complete kit for growing paper-white narcissuses, including bulbs, soil, pot, and saucer.* ● *Verpackung für Narzissenzwiebeln, Erde, Topf und Teller. Als Resultat seiner Bemühungen darf der Käufer auf weisse Narzissen hoffen.* ▲ *Narcisses blancs. Kit complet du parfait petit jardinier comprenant bulbes, terreau, pot et soucoupe.*

PAGE 127, *image 288* ART DIRECTORS/DESIGNERS: *Melanio R. Gomez, Darren Kearns, Mary Lui* PRODUCT PHOTOGRAPHER: *Shelby Burt* PHOTOGRAPHER: *Shelby Burt* AGENCY: *Top Spin Design* CLIENT: *Metropolitan Transit Authority (New York Transit Museum Gift Store)* COUNTRY: *USA* PRINTER: *Colonial Printers* TYPEFACE: *Goudy, Goudy Bold, Helvetica Bold* PAPER: *Fason-soft cream, crack-n-peel plus, #70 Premium uncoated* ■ *The goal was to create an industrial package to house*

a chrome-finish, hand-enameled, clip pen. An aluminum cannister contains a muslin drawstring bag which is pulled out of the cannister by an antique subway token. The label is printed steel grey and the subway train stop icons act as the pen clip. The border pattern is reminiscent of subway mosaics. This package was designed to be sold along with a line of subway theme products. ● *Diese industriell anmutende Alu-Verpackung enthält ein Säckchen mit einer Kordel, das sich mit Hilfe einer alten U-Bahn-Münze herausziehen lässt. Das darin enthaltene Schreibgerät hat eine Klammer, die aus Miniaturschildern von U-Bahnstationen besteht. Das Randmuster erinnert an die typischen U-Bahn-Mosaiks. Entworfen wurde die Verpackung für den Geschenkkiosk des NY Transit Museums, in dem eine ganze Reihe von Artikeln mit dem U-Bahn-Thema verkauft werden.* ▲ *Emballage industriel pour stylo clip chromé et laqué main. La boîte en fer blanc contient une petite bourse à cordon que l'on extrait avec un ancien jeton de métro. Le clip est formé de plaques de stations de métro miniatures. Le motif rappelle les mosaïques qui ornent les murs du métro new-yorkais. Ce packaging a été conçu pour la boutique-cadeaux du NY Transit Museum qui propose toute une série d'objets et de gadgets sur le thème du métro.*

PAGE 128, 129; *images 289–292* ART DIRECTOR: *Jun Sato* PRODUCT PHOTOGRAPHER: *Naoto Kato* AGENCY: *Jun Sato Design, Inc.* CLIENT: *Gallery Interform* COUNTRY: *Japan* PRINTER: *Kotobuki Seihan Insatsu Co., Ltd* TYPEFACE: *Frutiger Bold, Frutiger Regular, Universe Bold* PAPER: *Recycled Board* ■ *This packaging was designed to showcase the enclosed dispenser, to protect the product during distribution, to be recyclable and/or biodegradable, and to be low-cost.* ● *Diese Verpackung dient der Darstellung und dem Schutz des darin enthaltenen Produktes. Das Material musste wiederverwertbar und/oder biologisch abbaubar und dabei preisgünstig sein.* ▲ *L'emballage devait présenter le produit de manière séduisante tout en lui offrant une protection optimale. Le matériau devait être bon marché, recyclable et/ou biodégradable.*

PAGE 130, *image 293* ART DIRECTOR/DESIGNER/ARTIST/ILLUSTRATOR: *Tracy Holdeman* PRODUCT PHOTOGRAPHER: *Rock Island Studios* AGENCY: *Love Packaging Group* CLIENT: *The Hayes Company Inc.* PRINTER: *Love Box Company* TYPEFACE: *Opti Bernhard Bold Cursive, Copperplate 33BC, Senator Demi* PAPER: *Kraft Corrugated E-Floot* ■ *The packaging was intended to separate the product from the myriad of other standard cedar products in the home and garden market.* ● *Aufgabe dieser Verpackung war es, dem Produkt angesichts der Vielzahl von Zedernholzartikeln für den Haus- und Gartenbereich ein eigenständiges Image zu geben.* ▲ *Le packaging devait permettre au produit de se démarquer parmi la multitude d'articles en cèdre pour la maison et le jardin.*

PAGE 131, *image 294* ART DIRECTOR: *Robert Wood* DESIGN DIRECTOR: *Bill Capers* PROJECT LEADER: *Linda Harriman* DESIGNER: *Linda O'Neill, Lori Pilla, Joe Pozerycki, Mary Boisvert, Ellen Hartshorne, Carolina Senior, Brooks Beisch* PHOTOGRAPHER: *Peter Medilek* AGENCY: *Fitch Inc.* CLIENT: *Digital Equipment Corporation* COUNTRY: *USA* PRINTER: *Labels: Universal Press, Packaging: Advanced Design* TYPEFACE: *Helvetica* ■ *The goal was to develop an attractive, cost-effective collection of point-of-purchase materials for the "Starion" line. The project also entailed the development of the graphic language for the new retail packaging design and support materials for other introductory products.* ● *Aufgabe war die Herstellung einer attraktiven, kostengünstigen Kollektion von Verkaufshilfen für die «Starion--Linie». Dazu gehörten auch die Entwicklung einer graphischen Sprache für die neue Verpackung und Promotionsmaterial für andere neue Produkte der Marke.* ▲ *Il s'agissait de créer à moindres coûts une PLV attractive pour la gamme «Starion». Le projet comprenait également le développement d'un langage graphique pour le packaging et le matériel publicitaire d'autres nouveautés de la marque.*

PAGE 132, *images 295, 297* DESIGNER: *Anders Tørris Christensen* ARTIST/ILLUSTRATOR: *Kjell Nupen* AGENCY: *Christensen/Lund* CLIENT: *Jotun A/S* COUNTRY: *Norway* PRINTER: *Skanem Moss, Norway*

PAGE 132, *image 296* ART DIRECTOR: *Annette Harcus* DESIGNER: *Kristin Thieme* PRODUCT PHOTOGRAPHER: *Keith Arnold* ILLUSTRATOR: *Kristin Thieme* AGENCY: *Harcus Design* CLIENT: *Wattyl Australia* COUNTRY: *Australia* PRINTER: *William McKersie*

PAGE 133, *image 298* ART DIRECTOR/DESIGNER: *Steve Sandstrom* ARTIST/ILLUSTRATOR: *Bill Sanderson* AGENCY: *Sandstrom Design* CLIENT: *Leupold* COUNTRY: *USA* TYPEFACE: *Copperplate, Franklin Gothic, Matrix* ■ *These waterproof matches were produced as a promotional gift theme for an outdoor accessories company.* ● *Diese wasserresistenten Streichhölzer dienen als Promotionsartikel für einen Hersteller von Freizeitartikeln.* ▲ *Alumettes étanches. Cadeau publicitaire d'un fabricant d'articles de loisirs.*

PAGE 134, *image 299* ART DIRECTOR: *Jack Anderson* DESIGNERS: *Jack Anderson, Heidi Favour, John Anicker, David Bates* PHOTOGRAPHER: *Tom Collicott* PRODUCT PHOTOGRAPHER: *Tom McMackin* ARTIST/ILLUSTRATOR: *John Fretz* AGENCY: *Hornall Anderson Design Works, Inc* CLIENT: *OXO International* COUNTRY: *USA* PRINTER: *Rand Whitney* TYPEFACE: *Futura* PAPER: *Litho Laminate of Kraft* ■ *This packaging was created for a product in a new line of housewares.* ● *Verpackung für eine neue Linie von Haushaltsgeräten.* ▲ *Emballage d'une nouvelle gamme d'appareils ménagers.*

PAGE 135, *image 300* ART DIRECTOR: *Jack Anderson* DESIGNERS: *Jack Anderson, Heidi Favour, John Anicker, David Bates* PHOTOGRAPHER: *Studio 3* PRODUCT

PHOTOGRAPHER: *Tom McMackin* ARTIST/ILLUSTRATOR: *John Fretz* AGENCY: *Hornall Anderson Design Works, Inc.* CLIENT: *OXO International* COUNTRY: *USA* PRINTER: *Taiwan* TYPEFACE: *Futura* PAPER: *Dull-coated stock with satin varnish, laminated to corrugated board.* ■ *This packaging was created for a product in a new line of housewares.* ● *Verpackung für eine neue Linie von Haushaltsgeräten.* ▲ *Emballage d'une nouvelle gamme d'appareils ménagers.*

PAGE 135, *image 301* CREATIVE DIRECTORS: *Felipé Bascope, Jon Gothold* ART DIRECTOR: *Jeff Labbé*, DESIGNERS: *Jeff Labbé, Garrison Smet* PHOTOGRAPHER: *Kimball Hall* ILLUSTRATOR: *Loudvik Akopyan* WRITER: *Eric Springer* AGENCY: *dGWB Advertising* CLIENT: *Qualcomm Inc.* COUNTRY: *USA* TYPEFACE: *Bauhaus, Eurostyle, Bernhard, Courier* ■ *The objective was to launch the brand and develop non-traditional packaging for a cellular phone technology and product manufacturer.* ● *Eine unkonventionelle Verpackung und Lancierung einer neuen Marke für einen Hersteller von Handies.* ▲ *Packaging original créé pour le lancement d'une nouvelle marque de téléphones cellulaires.*

PAGE 135, *image 302* ART DIRECTOR/DESIGNER/ARTIST/ILLUSTRATOR: *Kobe* DESIGNERS: *Jeff Johnson, Alan Levsink* AGENCY: *Duffy Design* CLIENT: *Yakima* COUNTRY: *USA*

PAGE 135, *image 303* ART DIRECTOR/DESIGNER: *Michael Sieger* PRODUCT PHOTOGRAPHER: *Christian Richters* AGENCY: *Sieger Design Consulting GmbH* CLIENT: *RC Ritzenhoff Cristal GmbH* COUNTRY: *Germany* PRINTER: *self-adhesive label: Druckerei Schulte* TYPEFACE: *Garamond* ■ *The various designs of these beer glasses were created by designers and architects. The series comprises thirty designs and is meant for everyday use as well as to be collectibles and gift items.* ● *Die Dekors dieser Biergläser wurden von verschiedenen Designern und Architekten entworfen. Die Serie, die insgesamt 30 Dekors umfassen soll, ist sowohl für den täglichen Gebrauch als auch als Sammel-und Geschenkartikel gedacht.* ▲ *Divers architectes et designers ont créé les motifs de ces verres à bière. Les différentes pièces de cette série, qui sera déclinée en 30 variations, sont destinées ââ l'usage quotidien, peuvent être collectionnées ou offertes en cadeau.*

PAGE 136, *image 304* ART DIRECTOR: *David Ceradini* DESIGNER: *Natalie Jacobs* AGENCY: *Ceradini Design* CLIENT: *The Great Atlantic & Pacific Tea Co.* ■ *The design objective was to develop a clean, simple graphic system for a large range of different style paper plates.* ● *Hier ging es um die Entwicklung eines sauberen, einfachen graphischen Systems für ein grosses Sortiment verschiedener Papierteller.* ▲ *Il s'agissait de développer un système graphique simple et convaincant pour un vaste assortiment d'assiettes en papier.*

PAGE 136, *image 305* ART DIRECTOR/DESIGNER: *David Ceradini* ARTISTS/ILLUSTRATORS: *Pete Mueller, Tim Shannon* AGENCY: *Ceradini Design* CLIENT: *The Hartz Mountain Coroporation* ■ *The goal was to communicate that this is a technologically advanced flea collar with a premium price point.* ● *Hier wird dem Käufer durch die Verpackung mitgeteilt, dass er ein Flohhalsband erster Güte vor sich hat.* ▲ *Collier antipuces pour animaux domestiques. Ce qui se fait de mieux en la matière, indique l'emballage.*

PAGE 136, *image 306, 307* AGENCY: *Cato Design* CLIENT: *Tupperware Australia* COUNTRY: *Australia* PRINTER: *Rothfield Print Management* TYPEFACE: *Gill Sans* ■ *This packaging was created as part of the client's strategy to present their products in sets. This re-design resulted in a series of gift-style boxes.* ● *Die Neugestaltung der Verpackung für den Kunden, der seine Produkte immer als Sets anbietet, resultierte in einer Reihe von Schachteln mit Geschenkcharakter.* ▲ *Ce packaging – une série de boîtes-cadeau – reflète la stratégie du client qui propose toujours ses produits en sets.*

PAGE 136, *image 307* AGENCY: *Cato Design* CLIENT: *Tupperware Australia* COUNTRY: *Australia* PRINTER: *Rothfield Print Management* TYPEFACE: *Gill Sans* ■ *This packaging was created as part of the client's strategy to present their products in sets. This redesign resulted in a series of gift-style boxes.* ● *Die Neugestaltung der Verpackung für den Kunden, der seine Produkte immer als Sets anbietet, resultierte in einer Reihe von Schachteln mit Geschenkcharakter.* ▲ *Ce packaging – une série de boîtes-cadeau – reflète la stratégie du client qui propose toujours ses produits en sets.*

PAGE 136, *image 309* ART DIRECTOR: *Robin Hall* DESIGNER: *Ian Stokes* AGENCY: *Davies Hall* CLIENT: *Bouchon, Ltd.* COUNTRY: *England* PRINTER: *Gaffyne and Brown Ltd.* TYPEFACE: *Futura Light* PAPER: *black single faced board, flute; 350 Highland Demi* ■ *The goal was to produce structural packaging and graphics that reflect the quality of this luxury stopper in a contemporary style.* ● *Verpackung und Graphik sollten auf zeitgemässe Art die Qualität dieses luxuriösen Flaschenverschlusses reflektieren.* ▲ *Packaging et graphisme dans l'air du temps pour un luxueux bouchon.*

PAGE 136, *image 310* ART DIRECTOR: *Glenn Tutssel* DESIGNER: *Garrick Hamm* AGENCY: *Tutssels* CLIENT: *Boots* COUNTRY: *England* TYPEFACE: *Sabon* ■ *The simple double crossing of two labels creates a cross in this design for a medicated toilet tissue.* ● *Das durch die beiden Etiketten gebildete Kreuz ist Teil der Graphik für medizinisch behandeltes Hygienepapier* ▲ *La croix formée par le chevauchement des deux étiquettes est un élément graphique utilisé pour un papier hygiénique medical.*

PAGE 136, *image 311* DESIGNER: *Kazuyuki Fujisawa* PRODUCT PHOTOGRAPHER: *Shinichi Hoshikawa* AGENCY: *TCD Corporation* CLIENT: *Kokuyo Co., Ltd.* COUNTRY: *Japan* TYPEFACE: *Frutiger Roman* PAPER: *Coated* ■ *This packaging design was created for a series of drawing tools for professional use.* ● *Packungsgestaltung für Zeichenmaterial, das für Profis bestimmt ist.* ▲ *Packaging de fournitures d'art pour usage professionnel.*

PAGE 137, *images 312, 313* ART DIRECTOR: *Tassilo von Grolman* DESIGNER/ARTIST/ILLUSTRATOR: *Annette Rust* PRODUCT PHOTOGRAPHER: *Norbert Latocha* AGENCY: *Tassilo von Grolman Design* COUNTRY: *Germany* ■ *The product was created as a promotional giveaway. The little herb garden consists of eight cubes contained in a decorative box. The single cubes are designed so that individual text blocks can be placed on them; when one turns the cubes, become one coherent text.* ● *Das Produkt ist als Kundengeschenk bzw als Promotion gedacht. Dieser kleine Kräutergarten besteht aus acht Würfeln in einer dekorativen Box. Die einzelnen Würfel sind so gestaltet, dass darauf individuelle Textblöcke plaziert werden können, die durch Drehen der Würfel zu einem zusammenhängenden Text werden.* ▲ *Cadeau publicitaire. Ce petit jardin de fines herbes se compose de huit cubes présentés dans une boîte décorative. Les blocs de caractères figurant sur les cubes forment un texte cohérent lorsque l'on assemble correctement les cubes.*

PAGE 138, *image 314* ART DIRECTOR/DESIGNER: *Glenn Tutssel* PHOTOGRAPHER: *Andy Seymour* AGENCY: *Tutssels* CLIENT: *Philips Lighting* TYPEFACE: *Futura* ■ *This packaging was created for a new flourescent light.* ● *Packungsgestaltung für eine neue, fluoreszierende Glühbirne.* ▲ *Emballage d'une nouvelle ampoule fluorescente.*

PAGE 139, *image 315* ART DIRECTOR: *Keizo Matsui* DESIGNER: *Yuko Araki* PHOTOGRAPHER: *Masao Chiba* AGENCY: *Hundred Design, Inc. (in-house)* COUNTRY: *Japan* PRINTER: *Kotobuki Printing Co., Ltd.* ■ *These images were created for a calendar.* ● *Diese Bilder waren für einen Kalender bestimmt.* ▲ *Images destinées à un calendrier.*

PAGE 140, *image 316* ART DIRECTOR: *Rick Tesoro* DESIGNER: *Dave Wang* PHOTOGRAPHER: *Bart Gorin* AGENCY: *Parham Santana, Inc.* CLIENT: *M.H. Segan & Company* ■ *The objective was to establish a point of distinction and claim territory for this new line. The non-conformist philosophy of the client was utilized to create packaging with personality.* ● *Ziel dieser Verpackung war ein spezieller Auftritt, um der Produktlinie ihren Platz im Markt zu sichern. Die nonkonformistische Einstellung des Auftraggebers erleichterte die Aufgabe.* ▲ *Packaging créé pour une nouvelle ligne de produits qui entend se démarquer et s'imposer par son originalité.*

PAGE 141, *images 317, 318* ART DIRECTOR/DESIGNER: *Minoru Tabuchi* PHOTOGRAPHER: *Akinori Hasegawa* AGENCY: *Daiko Advertising, Inc.* CLIENT: *Fujitsu Tokushima Systems Engineering Ltd.* COUNTRY: *Japan*

PAGE 142, *image 319* ART DIRECTOR/DESIGNER: *Scott Mires, Tracy Sabin* AGENCY: *Mires Design Inc.* CLIENT: *LA Gear* COUNTRY: *USA* ■ *To create a point-of-purchase display for cartridges and batteries for the client's footwear line, it was essential to create a strong image that extended to various applications such as t-shirts and keychains as well.* ● *Ein Display- und Beleuchtungssystem für Schuhe, das sich auch für verschiedene andere Bereiche wie T-Shirts und Schlüsselanhänger anwenden lassen sollte.* ▲ *Présentoir et système d'éclairage pour des chaussures, également utilisables pour d'autres articles tels que t-shirts et porte-clefs.*

PAGE 143, *image 320* ART DIRECTOR: *Robert Wood* DESIGN DIRECTOR: *Bill Capers* PROJECT LEADER: *Linda Harriman* DESIGNERS: *Linda O'Neill, Lori Pilla, Joe Pozerycki, Mary Boisvert, Ellen Hartshorne, Carolina Senior, Brooks Beisch* PHOTOGRAPHER: *Peter Medilek* AGENCY: *Fitch Inc.* CLIENT: *Digital Equipment Corporation* COUNTRY: *USA* PRINTER: *Labels: Universal Press, Packaging; Advanced Design* TYPEFACE: *Helvetica* ■ *The purpose was to develop an attractive, cost-effective collection of point-of-purchase materials for the "Starion" line. The project also entailed the development of the graphic language for the new retail packaging design and support materials for other introductory products.* ● *Aufgabe war die Herstellung einer attraktiven, kostengünstigen Kollektion von Verkaufshilfen für die «Starion»-Linie. Dazu gehörte auch die Entwicklung einer graphischen Sprache für die neue Verpackung und Promotionsmaterial für andere neue Produkte der Marke.* ▲ *Il s'agissait de créer à moindres coûts une PLV attractive pour la gamme «Starion». Le projet comprenait également le développement d'un langage graphique pour le packaging et le matériel publicitaire d'autres nouveautés de la marque.*

PAGE 143, *image 321* ART DIRECTOR/DESIGNER: *Yasuo Tanaka* AGENCY: *Package Land Co, Ltd.* CLIENT: *Saiwai Co.* ■ *This store packaging was designed for simple impact.* ● *Bei dieser für den Einzelhandel entworfenen Verpackung ging es um eine starke, einfache Wirkung.* ▲ *Créer un impact simple et direct, tel était l'objectif de cet emballage.*

PAGE 143, *image 322* CREATIVE DIRECTOR: *Kent Hunter, Aubrey Balkind* DESIGNERS: *Brett Gerstenblatt, Kin Yuen* AGENCY: *Frankfurt Balkind Partners* CLIENT: *Pantone, Inc.* COUNTRY: *USA* PRINTER: *Pantone, Inc.* ■ *The object of this package design was to reinforce the client's tagline "The Power of Color."* ● *-Die Macht der Farbe», der Slogan des Auftraggebers, war Thema dieser Packungsgestaltung.* ▲ *Le slogan du client, -Le pouvoir de la couleur», était le thème de cet emballage.*

PAGE 144, *image 323* ART DIRECTOR: *Steve Coleman* DESIGNERS: *Chris De Lisen, Simon Macrae* AGENCY: *Elton Ward Design* CLIENT: *Caltex Oils Australia* COUNTRY: *Australia* ■ This two year packaging project was created for a relaunch of the client's oil range. ● Bei diesem zwei Jahre dauernden Verpackungsprojekt ging es um einen neuen Auftritt für die Ölprodukte des Kunden. ▲ Conditionnement créé dans le cadre d'un projet portant sur deux ans et réalisé pour relancer la gamme d'huiles du client.

PAGE 144, *image 324* ART DIRECTOR/DESIGNER: *Yasuo Tanaka* AGENCY: *Package Land Co, Ltd.* CLIENT: *Eikosha Co. Ltd.* ■ This packaging for a car air freshener presents a selection of four fragrances. ● In dieser Verpackung sind Auto-Raumparfüms in vier verschiedenen Duftnoten enthalten. ▲ Emballage d'un désodorisant pour voitures proposé en quatre fragrances.

PAGE 144, *image 325* ART DIRECTOR: *Harry Pearce* DESIGNERS: *Harry Pearce, Mark Diater* PHOTOGRAPHER: *Richard Foster* ARTIST/ILLUSTRATOR: *Roger Taylor* AGENCY: *Lippa Pearce Design* CLIENT: *Halford's Limited* TYPEFACE: *Franklin Gothic*

PAGE 144, 145; *images 326, 327* ART DIRECTOR: *Gil Maia* DESIGNERS: *Gil Maia, Osvaldo Silva* PHOTOGRAPHER: *J.P. Sotto Mayor* PRODUCT PHOTOGRAPHER: *Camera* ARTIST/ILLUSTRATOR: *Osvaldo Silva* AGENCY: *Gadesi* CLIENT: *Silampos* COUNTRY: *Portugal* PRINTER: *Grafica Calheiros* TYPEFACE: *Gill Sans, Helvetica* PAPER: *Microware e flute +/- 300g. m2, board 220 g. m2* ■ This set of colorful pressure cooker boxes created for a Portuguese kitchenware plant was designed to allow for efficient stockage and visual appeal. ● Dieser Satz farbenfroher Verpackungen für Dampfkochtöpfe eines portugiesischen Herstellers sollte optisch attraktiv sein und sich gut lagern lassen. ▲ Série d'emballages aux couleurs vives pour les autocuiseurs d'un fabricant portugais. Il s'agissait de combiner séduction et qualités de stockage optimales.

PAGE 146, *image 328* ART DIRECTOR: *Silvana Conzelmann* DESIGNER/PHOTOGRAPHER: *Jean-Jacques Schaffner* ARTIST/ILLUSTRATOR: *Silvana Conzelmann* AGENCY: *Schaffner & Conzelmann AG* CLIENT: *Ciba Agron Schweiz* COUNTRY: *Switzerland* TYPEFACE: *Lucida Sans* ■ Redesign of the plant chemicals of Ciba-Geiby. ● Überarbeitung der Verpackung für Pflanzenschutzmittel von Ciba-Geigy. ▲ Nouvel emballage de produits phytosanitaires Ciba-Geigy.

PAGE 147, *image 329* ART DIRECTOR: *Keith Steimel* DESIGNERS: *Paul McDowall, Keith Steimel* ARTIST/ILLUSTRATOR: *Paul McDowall* AGENCY: *Cornerstone* CLIENT: *Chebby Foods* COUNTRY: *USA* ■ This packaging for a topical analgesic was created to portray a high degree of efficiency. ● Bei dieser Verpackung für ein Schmerzmittel ging es um die Darstellung seiner schnellen Wirkung. ▲ Emballage d'un médicament topique mettant en exergue l'efficacité rapide du remède.

PAGE 147, *image 330* ART DIRECTOR/DESIGNER/TYPOGRAPHER: *Danny Klein* PRODUCT PHOTOGRAPHY: *Verenigde Bedrijven Fototechniek BV* ARTIST/ILLUSTRATOR: *Caroline Church* AGENCY: *Garden Studio, Milford-Van den Berg Design* CLIENT: *Sweetlife B.V.* PRINTER: *Danapak Flexibel a/s* ■ This design uses air as a symbol for the "breathy" effect of this throat pastil. ● Luft dient hier als Symbol für die befreiende Wirkung der Halstabletten. ▲ L'air, symbole de l'effet «rafraîchissant» et bienfaisant de cette pastille pour la gorge.

PAGE 148, *image 331* DESIGNERS: *Giovanni Pellone, Bridget Means* PHOTOGRAPHER: *Paul Managan* PRODUCT PHOTOGRAPHER: *Giovani Pellone* AGENCY: *Pellone & Means* CLIENT: *Manostat Corporation* COUNTRY: *USA* PRINTER: *Manpak* TYPEFACE: *Garamond* PAPER: *0.010 SBS, coated one side* ■ The tubing sample package was designed for a manufacturer of laboratory equipment. The shape allows the samples to be held and identified after they have been removed from the package. ● Diese Musterverpackung wurde für einen Hersteller von Laborausrüstungen entworfen. Die Form erlaubt eine sichere Verpackung und Aufbewahrung des Inhalts. ▲ Emballage d'échantillons en forme de tube conçu pour un fabricant de matériel de laboratoire. La forme permet de tenir et d'identifier les échantillons une fois sortis de l'emballage.

PAGE 149, *images 332–333* ART DIRECTOR: *Stefan Oevermann* DESIGNER: *Carolin Peiseler* ARTIST/ILLUSTRATOR: *Claudia Less* AGENCY: *PHARMA PERformANCE GmbH* CLIENT: *Mundipharma GmbH* COUNTRY: *Germany* PRINTER: *PHARMA PERformANCE* TYPEFACE: *Futura Bold* ■ For this "Power Man" campaign, an original Power-Pack containing Trumundin®-gimmicks was created. The slogan referring to the client as "pain specialist" draws attention to their broad dosage range of painkiller products. ● Packungsgestaltung im Rahmen einer speziellen Kampagne für ein Schmerzmittel. Es ging darum, den Hersteller als «Scherzspezialisten» darzustellen und auf sein breites Schmerzmittelsortiment hinzuweisen. ▲ Emballage utilisé dans le cadre d'une campagne publicitaire pour un remède contre la douleur. Il est fait allusion au fabricant, «Le spécialiste de la douleur», et à sa vaste gamme de produits.

PAGE 150, *image 334* ART DIRECTOR: *Doug Akagi* DESIGNER: *Carriet Worthen* ARTIST/ILLUSTRATOR: *John Hersey* AGENCY: *Akagi Remington* CLIENT: *Opcode Music Systems* COUNTRY: *USA* ■ This music software package was designed to stand out on a shelf and to explain all the features of the product. ● Bei dieser Verpackung für Musik-Software ging es um Produktinformation und die Wirkung im Verkaufsgestell. ▲ Packaging d'un logiciel de musique conçu pour se démarquer en linéaire tout en expliquant les spécificités du produit.

PAGE 150, *image 335* ART DIRECTOR: *Mary Scott* DESIGNERS: *Paul Farris, Mark Verlander, David Chapple* PHOTOGRAPHER: *Donald Miller* AGENCY: *Maddocks and Co.* CLIENT: *Sony Computer Entertainment* COUNTRY: *Japan* TYPEFACE: *Template Gothic, Template Gothic Bold, OCRB Symbol, Helvetica* ■ This game system packaging was created for the consumer electronics market targeting males ages 12—25. ● Diese Verpackung für Computerspiele richtet sich an ein männliches Zielpublikum im Alter von 12 bis 25 Jahren. ▲ L'emballage de ces jeux informatiques cible la tranche des 12–25 ans d'un public essentiellement masculin.

PAGE 150, *image 336* ART DIRECTOR/DESIGNER: *Hock Wah Yeo* DESIGNER: *Kelly Low* PRODUCT PHOTOGRAPHER: *Big Time Productions* AGENCY: *The Design Office of Wong & Yeo* CLIENT: *Digital Pictures* COUNTRY: *USA* PRINTER: *Everett Graphics* ■ This packaging was created for a fast-action Kung Fu computer game. ● Diese Verpackung enthält ein Kung-Fu-Computerspiel, in dem es um schnelle Reaktionen geht. ▲ Packaging d'un jeu informatique Kung Fu où la rapidité des réactions du joueur est mise à rude épreuve.

PAGE 150, *image 337 Double Switch Game* ART DIRECTOR: *Hock Wah Yeo* DESIGNERS: *Hock Wah Yeo, Kelly Low* PRODUCT PHOTOGRAPHY: *Big Time Productions* AGENCY: *The Design Office of Wong & Yeo* CLIENT: *Digital Pictures* COUNTRY: *USA* PRINTER: *Everett Graphics* ■ This packaging was created for a mystery adventure computer game. ● Packungsgestaltung für ein Abenteuer-Computerspiel. ▲ Packaging d'un jeu informatique d'aventures.

PAGE 151, *image 338* ART DIRECTOR: *David Jensen* DESIGNERS: *David Jensen, Mark Bird, Jennifer Hayes* PHOTOGRAPHER: *Tom Hollar* AGENCY: *Jensen Design Associates* CLIENT: *Canon Computer Systems, Inc.* PRINTER: *Calsonic Miura Graphics* ■ This package was distributed to the media by Canon. It is designed to contain all relevant product and company information. ● Dieses Programmpaket mit Informationen über die Firma und ihre Produkte wurde von Canon an die Medien versandt. ▲ Destiné aux médias, ce kit contient toutes les informations utiles sur la société Canon et ses produits.

PAGE 152, *image 339* ART DIRECTOR/DESIGNER: *Anthony Luk* ARTIST/ILLUSTRATOR: *Steven Lyons* AGENCY: *Profile Design* CLIENT: *Ascend Communications, Inc.* COUNTRY: *USA* PRINTER: *Citation Press* TYPEFACE: *Industria, Bank Gothic* PAPER: *22PT Springhill* ■ This packaging was created for a firm which develops, manufactures, and markets wide-area-network accessible products that support a large spectrum of end-user applications. ● Programmpaket einer Firma, die Produkte entwickelt, herstellt und vermarktet, die im Netzwerk zugänglich sind und ein grosses Spektrum von Anwender-Applikationen unterstützen. ▲ Packaging créé pour une société qui développe, fabrique et commercialise des produits accessibles sur un réseau très entendu, supportant une large gamme d'applications utilisateur.

PAGE 152, *image 340* ART DIRECTOR: *Lana Rigsby* DESIGNERS: *Lana Rigsby, Troy S. Ford* ILLUSTRATOR: *Andy Dearwater* AGENCY: *Rigsby Design* CLIENT: *Present Technologies* COUNTRY: *USA* ■ This videotape packaging was created for a company which provides presentation support equipment (projectors, monitors, lecterns, etc.) ● Diese Videokassettenverpackung wurde für eine Firma erstellt, die Projektoren, Monitoren und andere Geräte für Präsentationen verleiht. ▲ Cet emballage de vidéocassette a été conçu pour une société qui fournit des supports de présentation tels que projecteurs, moniteurs, etc.

PAGE 153, *image 341* ART DIRECTORS: *Tom Antista, Thomas Fairclough* PRODUCT PHOTOGRAPHY: *Michael West Photography* AGENCY: *Antista Fairclough Design* CLIENT: *Turner Home Entertainment* COUNTRY: *USA* ■ The package was developed for the distribution of promotional materials for the home video division of Turner Broadcasting. ● Packungsgestaltung für den Versand von Promotionsmaterial für Heim-Videos von Turner Broadcasting. ▲ Packaging destiné à l'envoi de matériel promotionnel pour le département vidéo domestique de Turner Broadcasting.

PAGE 154, *images 342–346* DESIGNER: *Tom Hough* ARTIST/ILLUSTRATORS: *Tom Hough, Steve Lyons* AGENCY: *Sibley Peteet Design* CLIENT: *Nortel* COUNTRY: *USA* PRINTER: *Wace/Chicago* ■ This package for a telecommunications equipment company is designed to hold approximately one year's worth of three publications put out by the company. ● Verpackung für Nortel, Hersteller von Geräten für die Telekommunikation. Sie bietet Platz für einen Jahrgang von drei Publikationen, die von der Firma herausgegeben werden. ▲ Packaging créé pour la société de télécommunications Nortel, et pouvant contenir la totalité des trois publications diffusées en une année par cette société.

PAGE 155, *image 347* ART DIRECTOR/DESIGNER: *Hock Wah Yeo* PRODUCT PHOTOGRAPHY: *Bigtime Productions* ARTIST/ILLUSTRATOR: *Rich Coben* AGENCY: *The Design Office of Hock Wah Yeo* CLIENT: *Spectrum Holobyte* COUNTRY: *USA* PRINTER: *Bertco* TYPEFACE: *Futura, Helvetica* ■ This package was designed to hold a CD-ROM computer game. ● Packungsgestaltung für ein CD-ROM Computerspiel. ▲ Packaging pour un jeu sur CD-ROM.

PAGE 156, *image 348* ART DIRECTOR/DESIGNER: *Anthony Luk* ARTIST/ILLUSTRATOR: *Steven Lyons* AGENCY: *Profile Design* CLIENT: *Ascend Communications, Inc.* COUNTRY: *USA* PRINTER: *Citation Press* TYPEFACE: *Industria, Bank Gothic* PAPER: *22PT Springhill* ■ This packaging was created for a firm which develops, manufactures, and markets wide-area-network accessible products that support a large spectrum of end-

user applications. ● *Packungsgestaltung für eine Firma, die Produkte entwickelt, herstellt und vermarktet, die im Netzwerk zugänglich sind und ein grosses Spektrum von Anwender-Applikationen unterstützen.* ▲ *Packaging créé pour une société qui développe, fabrique et commercialise des produits accessibles sur un réseau très étendu, supportant une large gamme d'applications utilisateur.*

PAGE 156, *image 349* ART DIRECTORS: *Thom Holden, Nelson Abbott* DESIGNERS: *Dave Bell, Rose Matusiak, Thom Holden* PHOTOGRAPHY: *Seymour Medick Photography* ARTIST/ILLUSTRATOR: *Dave Bell* AGENCY: *Thom & Dave Marketing Design* CLIENT: *Wharton Executive Education, University of Pennsylvania* COUNTRY: *USA* PRINTER: *CR Waldman, Jessie Jones Box Company* ■ *This packaging is intended to promote the philosophy that the only lifelong job is learning and that Wharton Executive Education promotes a link to data, information, knowledge, and wisdom in the world of business today.* ● *Der einzige Job auf Lebenszeit ist das Lernen, und Wharton Executive Education bietet der Geschäftswelt Zugang zu Daten, Informationen, Wissen und Weisheit.* ▲ *Le seul devoir est celui d'apprendre. Wharton Executive Education permet au monde des affaires d'accéder à des banques de données, à des informations, à la connaissance et à la sagesse.*

PAGE 157, *image 350* ART DIRECTOR/DESIGNER/ARTIST/ILLUSTRATOR: *David Rossi* PHOTOGRAPHER: *David McGrath* AGENCY: *The Hibbert Group* CLIENT: *Texas Instruments* COUNTRY: *USA* PRINTER: *RR Donnelly* TYPEFACE: *Giotto Bold* ■ *This packaging was designed for client/server software CD-Roms and manuals.* ● *Packungsgestaltung für Kunden/Server Software CD-ROMS und Handbücher.* ▲ *Packaging conçu pour des manuels et des CD-Roms client-serveur.*

PAGE 157, *image 351* ART DIRECTOR: *Rita Damore* DESIGNERS: *Rita Damore, Jackie Zaremba, Rob Mesarchik* PHOTOGRAPHER: *Mark Johann* ARTIST/ILLUSTRATOR: *Rob Mesarchik* AGENCY: *Damore-Johann Design* CLIENT: *CH Products* COUNTRY: *USA* TYPEFACE: *Industria, Universe Condensed* PAPER: *SBS 20 pt.* ■ *This packaging system for a flight simulation joystick was designed to illustrate the products while capturing the users' game play experience. The packaging was also designed to be easily printed in multiple languages; copy can be easily changed through black plate changes.* ● *Verpackungssystem für einen Flugsimulator-Joystick. Es ging um die Darstellung des Produktes und der Freude an diesem Spiel, wobei die Verpackung die Beschriftung in verschiedenen Sprachversionen vorsieht, die man durch einfachen Plattenwechsel nach Bedarf eindrucken kann.* ▲ *Système de packaging pour joystick de simulation de vol. Il s'agissait de présenter le produit tout en suscitant l'envie de jouer. L'emballage a aussi été conçu pour être facilement imprimé en plusieurs langues: seul le noir change.*

PAGE 158, *image 352* ART DIRECTOR: *Charles S. Anderson* DESIGNERS: *Charles S. Anderson, Joel Templin, Paul Howalt, Brian Smith, Tom Eslinger* ARTIST/ILLUSTRATOR: *CSA Archive* AGENCY: *CSA Archive (in-house)* COUNTRY: *USA* PRINTER: *KEA Incorporated* ■ *Interactive CD Packaging.* ● *Packungsgestaltung für eine interaktive CD.* ▲ *Boîtier d'un CD interactif.*

PAGE 159, *images 353, 354* ART DIRECTOR: *Jaimie Alexander* SENIOR DESIGNER: *Kate Murphy* DESIGNERS: *Paul Lycett, Kerry Larimer, Sarah Spatt, Joanie Hupp* PHOTOGRAPHER: *Mark Steele* AGENCY: *Fitch Inc.* CLIENT: *Iomega Corporation* COUNTRY: *USA* ■ *This packaging is part of a major corporate repositioning strategy to change the way the marketplace views, uses, and buys removable data storage products.* ● *Mit dieser Verpackung will der Hersteller die Einstellung, die Verwendung und die Kaufgewohnheiten im Hinblick auf Removable-Data-Storage-Produkte verändern.* ▲ *Cet packaging fait partie d'une importante stratégie de repositionnement del'entreprise, visant à changer la façon dont le marché considère, utilise et achète des produits de stockage de données amovibles.*

PAGE 159, *image 355* ART DIRECTOR: *Joseph Rattan* DESIGNERS: *Joseph Rattan, Greg Morgan* PHOTOGRAPHER: *Jay Brousseau (Image Bank Stock)* AGENCY: *Joseph Rattan Design* CLIENT: *The Image Bank* COUNTRY: *USA* ■ *This photo CD of selected stock images promotes an international stock photography source.* ● *Verpackung für eine Photo-CD mit Archiv-Photographie, die von einem internationalen Photoarchiv angeboten werden.* ▲ *Ce CD photo présentant une sélection d'images d'archives, sert à promouvoir une banque d'images internationale.*

PAGE 159, *image 356* ART DIRECTOR/ARTIST/ILLUSTRATOR: *Craig Frazier* DESIGNERS: *Craig Frazier, René Rosso* AGENCY: *Frazier Design* CLIENT: *Xaos Tools Inc.* COUNTRY: *USA* PRINTER: *Hatcher Trade Press* ■ *This packaging for graphics and animation effects software targets the creative user.* ● *Diese Verpackung für eine Software für graphische und Animations-Effekte wendet sich an Gestalter.* ▲ *Packaging d'un logiciel graphique et d'animation pour designers créatifs.*

PAGE 159, *image 357* ART DIRECTORS: *Primo Angeli, Rolando Rosles* DESIGNERS: *Phillippe Becker, Rolando Rosles* AGENCY: *Primo Angeli* CLIENT: *Verifone, Inc.* COUNTRY: *USA* ■ *The task was to develop a logo and sleek packaging to denote the speed of worldwide communications for software. The different packages in the series reflect a range of services.* ● *Hier ging es um das Logo und eine schnittige Verpackung für Software, die die Schnelligkeit der weltweiten Kommunikation zum Ausdruck bringen sollte. Die verschiedenen Verpackungen dieser Reihe stellen eine Reihe von Diensten dar.* ▲ *Eléments clés: le logo et packaging raffiné évoquant la vitesse de communication des logiciels à l'échelle mondiale. Les différents packagings de cette série présentent la gamme de services.*

PAGE 159, *image 358* ART DIRECTOR/DESIGNER: *Anthony Luk* ARTIST/ILLUSTRATOR: *Steven Lyons* AGENCY: *Profile Design* CLIENT: *First Virtual Corporation* COUNTRY: *USA* PRINTER: *Strategy Man* TYPEFACE: *Bank Gothic* PAPER: *22PT Springhill* ■ *This packaging was created for Media Operating Software, a product of a designer and manufacturer of multimedia applications for business users.* ● *Packungsgestaltung für Media-Betriebssoftware, ein Produkt eines Designers und Herstellers von Multimedia-Anwendungen für Unternehmen.* ▲ *Packaging créé pour Media Operating Software par un designer-fabricant d'applications multimédias, ciblant un public de professionnels.*

PAGE 159, *image 359* ART DIRECTOR: *Mike Hicks* DESIGNERS: *Shawn Harrington, Mick Hicks* PHOTOGRAPHER: *Kirk Tuck* ARTISTS/ILLUSTRATORS: *Shawn Harrington, Bill Geister, David Villareal* AGENCY: *HIXO, Inc.* CLIENT: *Relativity: a division of Liant Software* COUNTRY: *USA* TYPEFACE: *Helvetica Extended* ■ *The goal was to create a compelling visual package for new translation software that would create interest and push the visual envelope of the traditional DOS-based programming market.* ● *Das Ziel war eine optisch attraktive Verpackung für eine neue Übersetzungs-Software, mit der das Interesse des Verbrauchers geweckt und der visuelle Auftritt des traditionellen Programmiermarktes auf DOS-Basis verstärkt werden sollte.* ▲ *Le but était de créer un packaging attractif pour de nouveaux logiciels de traduction, capable d'éveiller l'intérêt du consommateur et de bousculer l'image du marché de la programmation DOS, bien traditionnelle.*

PAGE 159, *images 360, 361* ART DIRECTOR: *Julia LaPine* DESIGNERS: *Julia LaPine, Heidi Favour* PHOTOGRAPHER: *Kevin Cruff* PRODUCT PHOTOGRAPHER: *Tom McMackin* AGENCY: *Hornall Anderson Design Works, Inc.* CLIENT: *Megahertz Corporation* COUNTRY: *USA* PAPER: *Carton Stock.* ■ *This packaging was created for Megahertz software.* ● *Packungsgestaltung für Megahertz-Software.* ▲ *Packaging pour logiciel Megahertz.*

PAGE 159, *image 362* ART DIRECTOR/DESIGNER: *Anthony Luk* ARTIST/ILLUSTRATOR: *Steven Lyons* AGENCY: *Profile Design* CLIENT: *Ascend Communications, Inc.* COUNTRY: *USA* PRINTER: *Citation Press* TYPEFACE: *Industria, Bank Gothic* PAPER: *22PT Springhill* ■ *This packaging was created for a firm which develops, manufactures, and markets wide-area-network accessible products that support a large spectrum of end-user applications.* ● *Packungsgestaltung für eine Firma, die Produkte entwickelt, herstellt und vermarktet, die im Netzwerk zugänglich sind und ein grosses Spektrum von Anwender-Applikationen unterstützen.* ▲ *Packaging créé pour une société qui développe, fabrique et commercialise des produits accessibles sur un réseau très étendu, supportant une large gamme d'applications utilisateur.*

PAGE 159, *image 363* CREATIVE DIRECTOR: *John Burgess* ART DIRECTOR/ DESIGNER: *Fabian Schmid* PRODUCT PHOTOGRAPHY: *Jim Linna Photographics* ARTIST/ILLUSTRATOR: *Lisa Walker* AGENCY: *Werkhaus Design, Inc.* CLIENT: *Emagic, GmbH* COUNTRY: *Germany* TYPEFACE: *Caslon 540, Frutiger* ■ *A packaging system, logo designs, and digital illustration were created for an advanced, pro-audio software product. The system is designed to work across a number of different products, using color and logotype to differentiate the various software versions.* ● *Packungssystem, Logos und digitale Illustration für ein hochentwickeltes Pro-Audio-Software-Produkt. Die Verpackung musste sich für verschiedene Produkte eignen, wobei durch Farbkodierung und Logos zwischen den verschiedenen Software-Versionen unterschieden wird.* ▲ *Système de packaging, création de logos et d'images numériques pour un logiciel pro-audio high-tech. Le système devait être applicable à différents produits – couleurs et logos différenciant les versions.*

PAGE 159, *image 364* ART DIRECTOR: *Michael Osborne* DESIGNER: *Anna Shteerman* PRODUCT PHOTOGRAPHER: *Tony Stromberg* ARTIST/ILLUSTRATOR: *Steven Lyons* AGENCY: *Michael Osborne Design* CLIENT: *Paragraph International* COUNTRY: *USA* PRINTER: *Joyce Printing* TYPEFACE: *Block, Universe,OCRA, OCRB* ■ *Package design created for software that allows users to create 3-D multimedia galleries for the World Wide Web.* ● *Der Anwender dieser Software ist in der Lage, dreidimensionale Multimedia-Galerien für das World Wide Web herzustellen.* ▲ *Logiciel permettant aux utilisateurs de créer des galeries multimédia tridimensionnelles sur le site World Wide Web.*

PAGE 160, *image 365* ART DIRECTOR: *Greenville Main* ART DIRECTOR: *Diana Bidwell* DESIGNERS: *Diana Bidwell, Sarah Williams* PHOTOGRAPHER: *Mike Overeno* AGENCY: *BNA Design* CLIENT: *Backhouse Interiors* COUNTRY: *New Zealand* PRINTER: *Wellington Screen Print Ltd.* TYPEFACE: *Gill, Industria* PAPER: *White-back formacote* ■ *This sample case was created for product brochures.* ● *Für Produktbroschüren entworfene Box.* ▲ *Boîte pour des brochures de produits.*

PAGE 160, *image 366* ART DIRECTOR: *Michael Osborne* DESIGNER/ARTIST/ ILLUSTRATOR: *Kristen Clark* PRODUCT PHOTOGRAPHER: *Tony Stromberg* AGENCY: *Michael Osborne Design* CLIENT: *Cardinal Technologies, Inc* COUNTRY: *USA* PRINTER: *Joyce Printing* TYPEFACE: *Garamond* ■ *This packaging was designed for a line of modems positioned for the small business and home office.* ● *Packungsgestaltung für eine Reihe von Modems, die sich für die Anwendung in kleinen Unternehmen und Privathaushalten eignen.* ▲ *Packaging pour une gamme de modems destinées aux petites entreprises et aux particuliers.*

PAGE 160, *image 367* ART DIRECTORS/DESIGNERS: *Earl Gee, Fani Chung* PRODUCT PHOTOGRAPHER: *Kirk Amyx* ARTIST/ILLUSTRATOR: *Robert Pastrana* AGENCY: *Gee + Chung Design* CLIENT: *Xinet* COUNTRY: *USA* PRINTER: *Hatcher Trade Press* TYPEFACE:

Mona Lisa Solid, Caslon 540, OCRA PAPER: *Springhill 18 pt. SBS C1S* ∎ *The unique trapezoidal shape allows the product to stand out in a retail environment as well as on the end-user's shelf. The restaurant "servers" function as a metaphor for the product, which shares files between Macintosh and Unix computers.* ● *Die einzigartige Trapezform verleiht dem Produkt Eigenständigkeit im Verkaufsbereich und auch beim Endverbraucher. Die Restaurant-«Server» dienen als Metapher für die gemeinsame Verwaltung von Macintosh- und Unix Computer-Datensätzen.* ▲ *Originale, la forme trapézoïdale accroche l'œil sur le lieu de vente tout comme sur l'étagère de l'utilisateur final. Les couverts de service sont une métaphore pour ce produit de gestion de fichiers commun entre ordinateurs Macintosh et Unix.*

PAGE 160, *image 368* ART DIRECTOR/DESIGNER: *Hock Wah Yeo* DESIGNER: *Steven Fukuda* ARTIST/ILLUSTRATOR: *Tod Herman* PRODUCT PHOTOGRAPHY: *Big Time Productions* AGENCY: *The Design Office of Wong & Yeo* CLIENT: *Reality Bytes* COUNTRY: *USA* PRINTER: *Rundell Printing: Molded Fibre Technology* ∎ *This packaging was created for a network computer game set in a post-nuclear holocaust environment.* ● *Verpackung für ein vernetztes Computerspiel, das in einem Horrorszenario nach einer Atomkatastrophe stattfindet.* ▲ *Packaging d'un jeu pour réseau informatique, ayant pour toile de fond un monde post-nucléaire apocalyptique.*

PAGE 161, *image 369* ART DIRECTOR/DESIGNER/ARTIST/ILLUSTRATOR: *Neil Powell* PHOTOGRAPHER: *Hugh Kretschner* AGENCY: *Duffy Design*

PAGE 161, *image 370* ART DIRECTOR: *Lowell Williams* DESIGNER: *Bill Carson* AGENCY: *Pentagram Design, Inc.* CLIENT: *Computer Curriculum Corporation California* COUNTRY: *USA* PRINTER: *Lithocraft* TYPEFACE: *Bodoni* PAPER: *Corrugated Kraft Box* ∎ *This packaging was created for an educational software program developed for the "media generation."* ● *Verpackung für ein Lernprogramm, das für die «Media-Generation» entwickelt wurde.* ▲ *Packaging d'un logiciel éducatif destiné à la nouvelle «génération média».*

PAGE 162, 163; *images 371–373* ART DIRECTOR: *Masaya Yamaguchi, Jun Sato* DESIGNER: *Jun Sato* PRODUCT PHOTOGRAPHER: *Naoto Kato* AGENCY: *Gallery Interform* CLIENT: *Art Against AIDS Project* COUNTRY: *Japan* PRINTER: *Kotobuki Seihan Insatsu Co., Ltd* TYPEFACE: *Times Regular, Gill Sans Bold, Rotis Sanserif Bold* PAPER: *Recycled Board* ∎ *Art Against AIDS Japan is an event in which artists all over the world offer their work to raise funds for the American Foundation for AIDS Research. This package consists of postcards featuring the participating artists' work. The package is designed to be economical, portable, and "environment-friendly." The elastic band represents the confinement of AIDS.* ● *Art Against AIDS ist eine Aktion von Künstlern aus aller Welt, die Werke zur Verfügung stellen, deren Erlös einer amerikanischen Stiftung für AIDS-Forschung zugute kommt. Diese Verpackung enthält Postkarten, auf denen die zum Verkauf angebotenen Kunstwerke abgebildet sind. Sie sollte sparsam, handlich und umweltfreundlich sein. Das Gummiband symbolisiert das Eingesperrtsein der AIDS-Kranken.* ▲ *A l'origine d'Art Against AIDS Japan, des artistes du monde entier qui font don de leurs œuvres afin de réunir des fonds destinés à l'American Foundation for AIDS Research. Cet emballage contient des cartes postales illustrant les travaux des artistes. L'emballage est conçu pour être économique, portable et respectueux de l'environnement. L'élastique symbolise l'«emprisonnement» que vivent les personnes touchées par le sida.*

PAGE 164, *image 374* ART DIRECTORS: *Seymour Chwast, Samuel Antupit* DESIGNER: *Seymour Chwast* PRODUCTION MANAGEMENT/PACKAGING: *The Actualizers* AGENCY: *The Pushpin Group* CLIENT: *Harry N. Abrams* COUNTRY: *USA* PRINTER: *Diversified Graphics, Inc.* TYPEFACE: *Hand-lettering, News Gothic* PAPER: *James River Paper Corporation* ∎ *This package sports ready-to-assemble paper animals, from Apollo the dog to Lana Banana the monkey.* ● *Diese Verpackung enthält Papiertiere zum Zusammensetzen.* ▲ *Cet emballage contient des animaux en papier à assembler.*

PAGE 164, *images 375, 376* ART DIRECTOR: *Dana Arnett* DESIGNERS: *Curtis Schreiber, Joy Panos Stauber* AGENCY: *VSA Partners, Inc.* CLIENT: *Harpo Productions, Inc* COUNTRY: *USA* PRINTER: *The Etheridge Printing Co, Wace* TYPEFACE: *Caslon* PAPER: *Simpson Teton* ∎ *This affiliate kit was produced in limited quantities for stations that carry the Oprah Winfrey Show. The kit houses literature, video, and radio shots.* ● *Dieses Set, zu dem Literatur, Video und Radioaufnahmen gehören, wurde in beschränkter Anzahl für TV-Regionalsender produziert, die die Oprah Winfrey Talk-Show im Programm haben.* ▲ *Produit en quantités limitées pour des chaînes de télévision régionales, ce set contient des extraits radiophoniques, des vidéos et de la littérature.*

PAGE 165, *images 377, 378* ART DIRECTOR/DESIGNER: *Peter Schmid* PRODUCT PHOTOGRAPHER: *Robert Striegl* ARTIST/ILLUSTRATOR: *Thomas Paster* AGENCY/CLIENT: *Fölser + Schernhuber* COUNTRY: *Austria* PRINTER: *Esterman* TYPEFACE: *Hand-lettering, Gamma, Snell* PAPER: *Corolla Label* ∎ *Designed as a Christmas gift for agency clients, this package features wild Canadian salmon wrapped in a map of Canada and packed in a wooden crate.* ● *Als Weihnachtsgeschenk für die Kunden einer Werbeagentur konzipierte Verpackung. Die Holzkiste enthält kanadischen Wildlachs, eingewickelt in eine Landkarte von Kanada.* ▲ *Emballage conçu comme cadeau de Noël pour les clients d'une agence de publicité. La caisse en bois contient du saumon sauvage canadien enveloppé dans une carte du Canada.*

PAGE 166, *image 379* ART DIRECTOR: *Boyd Dupree* DESIGNER: *Jay Simmons* PRODUCT PHOTOGRAPHER: *Scott Campbell* ARTIST/ILLUSTRATOR: *Brent Hale* AGENCY: *Hanson Design* CLIENT: *Wattu Wear, Inc.* COUNTRY: *USA* PRINTER: *B&D Coatings, Inc.* ∎ *This packaging was commissioned to promote a line of nostalgia wear designs. The idea was to capitalize on the current trend in oil and gas collectibles by packaging designs in items commonly found at gas stations (shop rags and cardboard). The product was displayed in an oil rack and given to clients at an invitation only event.* ● *Hier ging es um die Promotion von Kleidung mit nostalgischen Motiven. Als Inspiration dienten typische Tankstellenartikel, die gegenwärtig im Trend liegen. Das Produkt wurde auf einem Ölregal ausgestellt und an Kunden als Einladung für eine Veranstaltung abgegeben.* ▲ *Cet emballage avait pour fonction de promouvoir une ligne de vêtements au caractère nostalgique. Inspiré de l'environnement des stations-service et des articles que l'on y propose, ce produit a été exposé sur un présentoir habituellement destiné à de l'huile pour voitures et remis aux clients sous forme d'invitation à une manifestation unique.*

PAGE 166, *image 380* ART DIRECTOR/DESIGNER: *Kobe* DESIGNER: *Alan Levsink* AGENCY: *Duffy Design* CLIENT: *Yakima* COUNTRY: *USA* ∎ *This packaging was designed to promote a national conference held to strengthen the relationship between Minnesota and Canadian ceramic artists.* ● *Hier ging es um Werbung für eine Tagung zur Förderung der Beziehungen zwischen Keramikkünstlern aus Minnesota und Kanada.* ▲ *Il s'agissait de faire de la publicité pour une conférence nationale visant à renforcer les relations entre céramistes du Canada et du Minnesota.*

PAGE 167, *image 381* ART DIRECTOR/DESIGNER: *Jack Anderson* DESIGNER: *David Bates* PRODUCT PHOTOGRAPHER: *Tom McMackin* AGENCY: *Hornall Anderson Design Works, Inc.* CLIENT: *Smith Sport Optics* COUNTRY: *USA* PRINTER: *Union Bay Label* TYPEFACE: *Gil Sans Extended* PAPER: *Durotone, Speckletone, chipboard* ∎ *This packaging was created for a line of sunglasses.* ● *Packungsgestaltung für eine Linie von Sonnenbrillen.* ▲ *Emballage créé pour une ligne de lunettes de soleil.*

PAGE 167, *image 382* CREATIVE DIRECTORS: *Kent Hunter, Aubrey Balkind, Cheryl Heller* DESIGNER: *Kin Yuen* AGENCY: *Frankfurt Balkind Partners* CLIENT: *Pantone, Inc.* COUNTRY: *USA* PRINTER: *Pantone, Inc.* ∎ *The object of this package design was to reinforce the client's tagline "The Power of Color."* ● *«Die Macht der Farbe», der Slogan des Auftraggebers, war Thema dieser Packungsgestaltung.* ▲ *Le slogan du client, «Le pouvoir de la couleur», était le thème de cet emballage.*

PAGE 168, *images 383, 384* ART DIRECTOR: *Dennis Merritt* PRODUCTION COORDINATORS: *Kathy Thielmann, Sharon Long, Cathleen Spacil* DESIGNER/ARTIST/ILLUSTRATOR: *Dennis Merritt* PHOTOGRAPHER: *Rodney Rascona* WRITER: *Steve Hutchinsons* AGENCY: *SHR Perceptual Management* COUNTRY: *USA* ∎ *This design was developed for a new business presentation.* ● *Hier ging es darum, Land Rover als Kunden zu gewinnen.* ▲ *Ce design a été développé pour une présentation lors de laquelle il s'agissait d'acquérir un nouveau client.*

PAGE 169, *images 385, 386* ART DIRECTOR/DESIGNER: *Peter King Robbins* PRODUCT PHOTOGRAPHER: *Jeremy Samuelson* ARTIST/ILLUSTRATOR: *Peter King Robbins* AGENCY: *BRD Design (in-house)* COUNTRY: *USA* PRINTER: *Foundation Press Santa Monica, CA* FOUNDRY: *Ascast* TYPEFACE: *Arbitrary* PAPER: *French Aged Newsprint* ∎ *This self-promotion/holiday gift was created for a graphic design studio.* ● *Weihnachtsgeschenk als Eigenwerbung eines Graphik-Design-Studios.* ▲ *Ce cadeau de Noël, également utilisé à des fins promotionnelles, a été créé pour une agence de design graphique.*

PAGE 170, *image 387* ART DIRECTOR: *Akio Okumura* DESIGNER: *Katsuji Minami* AGENCY: *Packaging Create Inc.* CLIENT: *New Oji Paper Co., Ltd.* COUNTRY: *Japan* TYPEFACE: *Original*

PAGE 171, *images 388, 389* ART DIRECTOR/ DESIGNER: *Laura Gillespie* PRODUCT PHOTOGRAPHY: *Lang Photography* ILLUSTRATOR: *Chris Gall* AGENCY: *Northlich Stolley Lawarre Design Group* CLIENT: *Mead Communication Papers* COUNTRY: *USA* PRINTER: *Mead Carton* TYPEFACE: *Schneidler, Berling, Hand-drawn, CMY, Helevetica* ∎ *This packaging system was designed to work with all of the client's products. Each product has a carton and ream-wrapper package. The bottom of the package remains consistent throughout the line of products. The lid graphics change according to the brand.* ● *Dieses Verpackungssystem sollte sich für alle Produkte des Kunden eignen, was bedeutet, dass jeweils nur die graphische Gestaltung des Deckels für die verschiedenen Produktlinien verändert wird.* ▲ *Ce système d'emballage devait pouvoir être utilisé pour tous les produits du client, c'est-à-dire que seule la conception graphique du couvercle est changée en fonction de la ligne de produits.*

PAGE 172, *image 390* ART DIRECTOR/DESIGNER: *Rick Vaughn* PHOTOGRAPHER: *David Nufer* PRODUCT PHOTOGRAPHER: *Michael Barley* ARTIST/ILLUSTRATOR: *Rick Vaughn* AGENCY/CLIENT: *Vaughn Wedeen Creative* COUNTRY: *USA* PRINTER: *Albuquerque Printing* PAPER: *Simpson Starwhite Vicksburg* ∎ *This annual Christmas promotion was created to thank the people who helped the design firm throughout the year.* ● *Diese jährliche Weihnachtsgabe geht als Dank an die Kunden einer Design-Firma.* ▲ *Promotion de Noël annuelle créée pour une agence de design qui voulait ainsi remercier les clients et les personnes qui l'ont aidée tout au long de l'année.*

PAGE 173, *images 391–394* ART DIRECTORS/DESIGNERS: *Rob Verhaart, Ron Van der Vlugt* STYLIST: *Angelique Van Dam* PRODUCT PHOTOGRAPHER: *Arno Bosdma* AGENCY:

Designers Company CLIENT: *Hooghoudt Distillers B.V.* COUNTRY: *The Netherlands* ■ *This range of packaging was developed in conjunction with an event-campaign. The attire of the bottle is based on the imaginary associations from the history of the European royal houses (Louis the XIV, Marquis de Sade, the Church, the dungeon and the executioner). They are extreme characters who are internationally known and also fit with the contemporary party scenery. These promotional bottles were produced in small quantities and used for sales promotions.* ● *In kleinen Mengen hergestellte Verkaufspromotionen für Royalty-Wodka, die bei speziellen Anlässen und auch in Bars etc. ausgestellt werden. Thema ist die Geschichte der Königshäuser Europas, d.h. extreme Persönlichkeiten, die Kirche, Kerker und Henker etc.* ▲ *Bouteilles produites en petites quantités pour promouvoir la Royalty Vodka et destinées à des occasions spéciales. Le thème central est l'histoire des maisons royales d'Europe et les personnes ou images que l'on y associe, telles que Louis XIV, le Marquis de Sade, l'Eglise, le cachot et le bourreau.*

PAGE 174, *image 395* ART DIRECTOR: *Steve Wedeen* DESIGNERS: *Steve Wedeen, Adabel Kaskiewicz, Lucy Hitchcock, Dan Flynn* ARTIST/ILLUSTRATOR: *Vivian Harder* AGENCY: *Vaughn Wedeen Creative* CLIENT: *US West Communications* COUNTRY: *USA* PRINTER: *The Graphics Group* TYPEFACE: *Huxley Verticle* PAPER: *Monadbock Astrolite, Mohawk Superfine* ■ *This packaged internal sales promotion was designed to sell a telephone service. The package included prizes, apparel, and posters.* ● *Hier ging es um eine interne Promotionsaktion, die als Anreiz für das Verkaufspersonal beim Verkauf eines Telephondienstes gedacht war. Der Inhalt bestand aus Preisen, Kleidung und Plakaten.* ▲ *Promotion interne destinée au personnel de vente afin d'encourager les collaborateurs à vendre un service téléphonique. L'emballage incluait des prix, des vêtements et des affiches.*

PAGE 174, *image 396* ART DIRECTOR/DESIGNER: *Michael Sieger* PRODUCT PHOTOGRAPHER: *Christian Richters* AGENCY: *Sieger Design Consulting GmbH* CLIENT: *RC Ritzenhoff Cristal GmbH* COUNTRY: *Germany* PRINTER: *Self-adhesive label: Druckerei Schulte* TYPEFACE: *Garamond* ■ *The various designs of these beer glasses were created by designers and architects. The series comprises thirty designs and is meant for everyday use as well as to be used as collectors and gift items.* ● *Die Dekors dieser Biergläser wurden von verschiedenen Designern und Architekten entworfen. Die Serie, die insgesamt 30 Dekors umfassen soll, ist sowohl für den täglichen Gebrauch als auch als Sammel- und Geschenkartikel gedacht.* ▲ *Divers architectes et designers ont créé les motifs de ces verres à bière. Les différentes pièces de cette série, qui sera déclinée en 30 variations, sont destinées àâ l'usage quotidien, peuvent être collectionnées ou offfertes en cadeau.*

PAGE 174, *image 397* ART DIRECTOR/DESIGNER: *José Serrano* ARTIST/ILLUSTRATOR: *Nancy Stahl* PHOTOGRAPHER: *Carl Vandershuit* AGENCY: *Mires Design Inc.* CLIENT: *Deleo Clay Tile Company* COUNTRY: *USA*

PAGE 174, *image 398* ART DIRECTOR/DESIGNER: *José Serrano* ARTIST/ILLUSTRATOR: *Tracy Sabin* AGENCY: *Mires Design Inc.* CLIENT: *Cranford Street* COUNTRY: *USA* ■ *The Horoscope Calendar gives a different horoscope daily for every zodiac sign, as well as a full-page horoscope for birthdays.* ● *Dieser Kalender enthält für jedes Sternzeichen ein tägliches Horoskop sowie jeweils ein ganzseitiges Horoskop für die an dem Tag Geborenen.* ▲ *Ce calendrier présente pour chaque signe du zodiaque un horoscope journalier ainsi qu'un horoscope pleine page pour les dates d'anniversaire.*

PAGE 175, *image 399* ART DIRECTOR/DESIGNER/ARTIST/ILLUSTRATOR: *Sonia Greteman* AGENCY: *Greteman Group* CLIENT: *Hays Co.* COUNTRY: *USA* PRINTER: *Rand Graphics*

PAGE 176, *image 400* ART DIRECTOR/DESIGNER: *Glenn Tutssel* ARTIST/ILLUSTRATOR: *Marcus Jones* AGENCY: *Tutssels (in-house)* COUNTRY: *England* TYPEFACE: *Futura* ■ *This self-promotional mail shot was designed to promote a creative consultancy.* ● *Hier ging es um Werbung für eine Designfirma.* ▲ *Mailing autopromotionnel d'une agence de design.*

PAGE 177, *image 401* ART DIRECTOR: *Tor Hovind* DESIGNER: *Dave Gath* PRODUCT PHOTOGRAPHER: *Greg Page (Page One Studio)* AGENCY: *California State University, Long Beach* CLIENT: *National Paperbox Association* COUNTRY: *USA* ■ *This promotional packaging, designed as a student comp, includes the credit card and two gifts, a pre-paid calling card and a pocket phone book.* ● *Die an Studenten gerichtete Promotion besteht aus einer Kreditkarte und zwei Geschenken, einer Telephonkarte und einem kleinen Adressbuch.* ▲ *Cet emballage promotionnel destiné à des étudiants comprend une carte de crédit et deux cadeaux, à savoir une carte de téléphone et un carnet d'adresses.*

PAGE 177, *image 402* ART DIRECTOR: *Tor Hovind* DESIGNER: *Patricia Murphy* PRODUCT PHOTOGRAPHER: *Greg Page (Page One Studio)* AGENCY: *California State University, Long Beach* CLIENT: *National Paperbox Association* COUNTRY: *USA* ■ *This promotional packaging, designed as a student comp, is for distribution to corporate executives. Promotion includes two free gifts, a pen, and a business card case.* ● *Diese Promotion ist für Manager bestimmt und enthält zwei Geschenke, ein Schreibgerät und eine Schachtel für Visitenkarten.* ▲ *Cet emballage promotionnel s'adresse à des cadres et contient deux cadeaux: un stylo et un boîtier pour des cartes de visite.*

PAGE 177, *image 403* ART DIRECTOR: *Fabio Ongarato* DESIGNERS: *Ronnen Goren, Tim Richardson* PRODUCT PHOTOGRAPHER: *Jack Saraflan* AGENCY: *Fabio Ongarato*

Design CLIENT: *Andrew Simmons* COUNTRY: *Australia* TYPEFACE: *Helvetica Ultra Compresses, Helvetica Neue* ■ *This artist's promotional package features materials—a wooden box and papers—which are sympathetic to the artist's style of work.* ● *Diese Werbung eines Künstlers besteht aus einer Holzschachtel mit Papier, Material, das seinem Stil entspricht.* ▲ *Emballage promotionnel pour un artiste comprenant une boîte en bois et du papier, lesquels sont représentatifs de son style.*

PAGE 177, *image 404* ART DIRECTOR/DESIGNER: *Thuy Vuong* AGENCY: *Avancé Designs Inc.* CLIENT: *Campaign Inc.* COUNTRY: *USA* ■ *This packaging was designed to promote the client as a specialized, quality leather goods company.* ● *Hier ging es um die Präsentation des Kunden als Hersteller von speziellen Lederartikeln erster Qualität.* ▲ *Il s'agissait de présenter le client comme un fabricant d'articles en cuir de première qualité.*

PAGE 177, *image 405* ART DIRECTOR: *Nan Finkenaur* DESIGNER: *Carmine Vecchio* PHOTOGRAPHY: *Hotshots* ARTIST/ILLUSTRATOR: *Erik Ela* AGENCY: *Milton Bradley (in-house)*

PAGE 177, *image 406* ART DIRECTOR/DESIGNER: *Scott Mires* ARTIST/ILLUSTRATOR: *Tracy Sabin* AGENCY: *Mires Design Inc.* CLIENT: *Sunrise Publications* COUNTRY: *USA* ■ *The illustration and design of this packaging for a holiday edition of greeting cards, intend to convey a quality, handcrafted look.* ● *Illustration und Gestaltung dieser Verpackung für Grusskarten zu Feiertagen sollten den Eindruck eines hochwertigen, handgearbeiteten Produktes vermitteln.* ▲ *Deux critères ont prévalu à l'illustration et au design de cet emballage pour des cartes de vœux: donner l'impression que le produit est de première qualité et qu'il a été fait main.*

PAGE 177, *image 407* ART DIRECTOR: *Steve Wedeen* DESIGNERS: *Steve Wedeen, Vivian Harder, Lucy Hitchcock* ARTIST/ILLUSTRATOR: *Vivian Harder* AGENCY: *Vaughn Wedeen Creative* CLIENT: *US West Communications* COUNTRY: *USA* PRINTER: *The Graphics Group* TYPEFACE: *Futura, Koloss, Coronet* PAPER: *Monadbock Astrolite* ■ *This internal sales promotion was designed to incent sales force to sell a telephone service. The package included prizes, apparel, and posters; the "Lucky Lil" theme offered sales representatives chances to win "western" prizes, including a trip to a dude ranch in Wyoming.* ● *Eine interne Promotionsaktion als Anreiz für das Verkaufspersonal beim Verkauf eines Telephondienstes. Der Inhalt bestand aus Preisen, Kleidung und Plakaten. Das „Lucky Lil"-Thema bedeutete für die Vertreter die Chance, „Western"-Preise zu gewinnen, darunter eine Reise zu einer Touristen-Ranch in Wyoming.* ▲ *Promotion interne destinée au personnel de vente afin d'encourager les collaborateurs à vendre un service téléphonique. L'emballage incluait des prix, des vêtements et des affiches. Les représentants de commerce avaient également la possibilité de gagner des prix «western», dont un séjour dans un ranch touristique du Wyoming.*

PAGE 177, *image 408* ART DIRECTORS: *Steve Wedeen, Dan Flynn* DESIGNER: *Dan Flynn* PHOTOGRAPHER: *Valerie Santagto* ARTIST/ILLUSTRATOR: *Vivian Harder* AGENCY: *Vaughn Wedeen Creative* CLIENT: *US West Communications* COUNTRY: *USA* PRINTER: *The Graphics Group* TYPEFACE: *Industria Solid* PAPER: *Monadbock Astrolite* ■ *This packaged internal sales promotion was designed to incent sales force to a telephone service. The package included prizes, apparel, and posters; prizes included televisions, stereos, and other items that "turned on."* ● *Eine interne Promotionsaktion als Anreiz für das Verkaufspersonal beim Verkauf eines Telephondienstes. Der Inhalt bestand aus Preisen, Kleidung und Plakaten. Es gab u.a. Fernseh- und Stereogeräte sowie andere Dinge zu gewinnen, die den Besitzer 'anturnen'.* ▲ *Promotion interne destinée au personnel de vente afin d'encourager les collaborateurs à vendre un service téléphonique. L'emballage incluait des prix, des vêtements et des affiches. Parmi les prix à gagner: des téléviseurs, des chaînes stéréo et d'autres articles du même type.*

PAGE 178, *image 409* ART DIRECTOR/DESIGNER/PHOTOGRAPHER: *José Serrano* ARTIST/ILLUSTRATOR: *Tracy Sabin* AGENCY: *Mires Design Inc.* CLIENT: *Cranford Street* COUNTRY: *USA* ■ *The Talking Horoscope Alarm Clock gives a daily personal horoscope.* ● *Von diesem sprechenden Horoskop-Wecker erfährt der Besitzer sein tägliches Horoskop.* ▲ *Réveil avec horoscope parlant pour savoir de quoi chaque jour sera fait.*

PAGE 178, *image 410* ART DIRECTOR/DESIGNER/ARTIST/ILLUSTRATOR: *Thomas Vasquez* PRODUCT PHOTOGRAPHER: *Scott Metcalfe* AGENCY: *Squires & Company* COUNTRY: *USA* PRINTER: *Carich Printing* ■ *This party invitation in the form of a wine label was sent to the firm's clients.* ● *Diese Party-Einladung in Form eines Weinetiketts wurde an die Kunden der Firma verschickt.* ▲ *Carton d'invitation en forme d'étiquette envoyé aux clients de la société.*

PAGE 178, *image 411* ART DIRECTOR/DESIGNER/ARTIST/ILLUSTRATOR: *Sonia Greteman* PHOTOGRAPHER: *Paul Chauncy* AGENCY: *Greteman Group* CLIENT: *Kansas Lean* COUNTRY: *USA* PRINTER: *Rand Graphics* PAPER: *Litho label* ■ *This game was distributed by the public school system to educate children about the food pyramid and proper nutrition.* ● *Dieses Spiel wurde von den öffentlichen Schulen an die Kinder verteilt, um sie über richtige Ernährung aufzuklären.* ▲ *Ce jeu a été distribué à des enfants dans des écoles publiques afin de leur faire découvrir les secrets d'une alimentation saine et équilibrée.*

PAGE 178, *image 412* ART DIRECTOR: *Keith Steimel* DESIGNERS: *Paul McDowall, Keith Steimel* ARTIST/ILLUSTRATOR: *Tom Montini* AGENCY: *Cornerstone* CLIENT: *Moonlight Tobacco Co.* COUNTRY: *USA* ■ *The task was to create a unique umbrella* identity for a tobacco company. ● Aufgabe war die Schaffung eines einzigartigen, vielseitig anwendbaren Erscheinungsbildes für eine Tabakfirma. ▲ L'objectif était de créer une identité unique pour une entreprise de tabac.

PAGE 179, image 413 ART DIRECTOR: Stephanie Tinz DESIGNERS: Bernhard H. Tinz, Gerhard Eskuche ILLUSTRATORS: Barbara Simminger, Gerhard Eskuche AGENCY: Tinz. DCC CLIENT: M. Kaindl Holzindustrie COUNTRY: Germany ■ This product packaging was designed to effectively present chipboard samples. The result is a package that saves on material resources, is low-cost, environment-friendly, and distinct in its appearance. ● Die Dose, die in einer Papierrolle versandt wird, dient der Präsentation von Spanplattenmustern. Sie sorgt nicht nur für die Unverwechselbarkeit des Produktes und der Marke, sondern ist dank sparsamer Verwendung von Materialressourcen auch ökonomisch und ökologisch vertretbar. ▲ Ce packaging, emballé dans un rouleau de papier, sert à présenter des échantillons de plaques de fixation. Respectueux de l'environnement, il sert à identifier clairement le produit, nécessite peu de matériaux d'où son prix de revient bon marché.

PAGE 180, image 414 ART DIRECTOR/DESIGNER/ILLUSTRATOR: Sibylle Haase AGENCY: Atelier Haase & Knels CLIENT: Boutique Evelyn COUNTRY: Germany PRINTER: CABAS Verpackung & Design OHG ■ Design of a shopping bag for a fashion boutique. ● Gestaltung einer Einkaufstasche für eine Modeboutique. ▲ Sac pour une boutique de mode.

PAGE 181, images 415, 416 ART DIRECTOR/DESIGNER: Yasuo Tanaka CLIENT: Package Land Co., Ltd. ■ 3-D gift bag. ● Eine für Geschenke bestimmte Tragtasche. ▲ Pochette-cadeau tridimensionnelle.

PAGE 182, image 417 ART DIRECTOR: Akio Okumura DESIGNER: Katsuji Minami AGENCY: Packaging Create Inc. CLIENT: New Oji Paper Co., Ltd. COUNTRY: Japan TYPEFACE: Original

PAGE 183, image 418 ART DIRECTOR: Doo H. Kim DESIGNERS: Dongil Lee, Seunghee Lee AGENCY: DooKim Design CLIENT: Laneige, Pacific Group COUNTRY: Korea ■ This bag was designed for Utoo Zone, a fashion department store in the core of the fashion district of Seoul. "Have a nice day, you too!," the client's slogan, is utilized. ● Tragtasche für Utoo Zone, ein Mode-Kaufhaus im Herzen des Modedistrikts von Seoul, mit dem Slogan des Kunden: «Einen schönen Tag – auch für Sie !» ▲ Sac conçu pour Utoo Zone, un grand magasin situé au cœur du quartier de la mode à Séoul, avec pour slogan: «Bonne Journée, à vous aussi!»

PAGE 184, image 419 ART DIRECTOR: Kai Mui DESIGNERS: Kai Mui, Brian Johnson AGENCY: Mui + Gray CLIENT: Aki S.P.A. COUNTRY: Italy ■ The first logo symbolizes the power of winning and was featured in packaging and collateral used to introduce an Italian sportswear brand. The shopping bags and counter cards, available in Italy only, were given to all the retailers that carried the brand. ● Das erste Logo symbolisiert das Gefühl des Sieges und wurde für die Verpackung und anderes Werbematerial verwendet, um eine italienische Marke im Sportkleidungssektor einzuführen. ▲ Utilisé pour l'emballage et d'autres articles publicitaires afin de lancer une marque italienne de vêtements de sport, le premier logo symbolise le sentiment d'avoir remporté une victoire.

PAGE 185, image 420 ART DIRECTOR: Alan Chan DESIGNERS: Alan Chan, Peter Lo AGENCY: Alan Chan Design Company CLIENT: Kosta Boda COUNTRY: Hong Kong ■ Kosta Boda is a renowned Swedish manufacturer of high quality glassware. The multi-color, overlapping letters of the client's name enhance the principal characteristics of the product. ● Kosta Boda ist eine bekannte schwedische Marke für qualitativ hochwertige Glasartikel. Die mehrfarbigen, überlappenden Buchstaben des Namens des Herstellers entsprechen den Haupteigenschaften des Produktes. ▲ Kosta Boda est un fabricant suédois réputé pour ses articles en verre. Les lettres multicolores superposées qui reprennent le nom du client font ressortir les principales caractéristiques du produit.

PAGE 186, image 421 ART DIRECTOR: Carol Beuttner DESIGNERS: Carol Beuttner, Victoria Stamm PRODUCT PHOTOGRAPHER: Richard Pierce CLIENT: Aramis Inc. COUNTRY: USA ■ This Christmas presentation was created for retail department store counters. ● Laden-Display für die Weihnachtszeit. ▲ Présentoir utilisé à l'occasion des fêtes de Noël et destiné aux grands magasins.

PAGE 187, image 422 ART DIRECTOR: Lowell Williams DESIGNERS: Lowell Williams, Bill Carson, Melinda Maniscalco ART/ILLUSTRATIONS: Century Italian painters AGENCY: Pentagram Design CLIENT: Gianfranco Lotti COUNTRY: USA TYPEFACE: Bodoni, Futura ■ This full range of packaging—bags, boxes, etc.—was created for a new line of leather products designed and produced in Italy. ● Die aus Tragtaschen, Schachteln etc. bestehende Linie wurde für ein neues Lederwarensortiment entwickelt, das in Italien entworfen und hergestellt wird. ▲ Cette famille d'emballages – sacs, boîtes, etc. – a été créée pour une nouvelle ligne d'articles en cuir conçus et fabriqués en Italie.

PAGE 187, image 423 CREATIVE DIRECTOR: Amy Knapp EXECUTIVE CREATIVE DIRECTOR: Peter Allen DESIGNER: Albert Treskin ILLUSTRATOR: Daniel Pelavin CLIENT: DFS Merchandising Ltd. COUNTRY: Hong Kong PRINTER: Sun's Paper Bags Hong Kong TYPEFACE: "Canton Market" (custom typeface) ■ This packaging concept was developed for Canton Market and includes souvenir packaging, shopping bag, box, label and hang tag. ● Zu diesem für 'Canton Market' entwickelten

Verpackungskonzept gehören Souvenir-Verpackungen, Tragtaschen, Schachteln, Etiketten und Anhänger. ▲ Développé pour Canton Market, ce concept d'emballage comprend des sacs, des boîtes, des étiquettes et des emballages-souvenir.

PAGE 188, image 424 ART DIRECTOR/DESIGNER: Sandra Haeberli PRODUCT PHOTOGRAPHER: Rafael Koller AGENCY: Sasart Grafik & Design Küsnacht CLIENT: N.A.P. Production COUNTRY: Switzerland PRINTER: Byland & CIE.AG TYPEFACE: Copperplate, Shelley Allegro Script PAPER: Tintoretto Melange 200gm Cashmere ■ S.E.A. is a company which sells products that are associated with water. Shown here is a project for a new bag and soap. ● S.E.A. ist eine Firma, die Produkte verkauft, die auf irgendeine Weise mit Wasser zu tun haben. Gezeigt ist ein Entwurf für eine neue Tragtasche und Seife. ▲ L'entreprise S.E.A. vend toutes sortes de produits qui ont un rapport avec l'eau. Ici, un projet pour un nouveau sac et un savon.

PAGE 189, image 425 DESIGNERS: Jack Anderson, Lisa Cerveny, Suzanne Haddon PRODUCT PHOTOGRAPHER: Tom McMackin AGENCY: Hornall Anderson Design Works, Inc. CLIENT: Jamba Juice COUNTRY: USA PRINTER: Lithographics TYPEFACE: Meta, Bembo PAPER: Kraft ■ Jamba Juice sells fresh juices and smoothies and needed a complete packaging system to represent their new identity. ● Jamba Juice verkauft frische Säfte und Frucht-Shakes und brauchte ein komplettes Verpackungssystem, das die neue Identität der Firma reflektieren sollte. ▲ Jamba Juice vend des jus de fruit frais et des boissons rafraîchissantes onctueuses. La société avait besoin d'une gamme complète d'emballages pour présenter sa nouvelle identité.

PAGE 189, image 426 ART DIRECTOR/DESIGNER: Angelo Sganzerla ARTIST/ILLUSTRATOR: Franco Testa AGENCY: Angelo Sganzerla CLIENT: L'Erbaolario ■ This shopping bag was created for a producer and vendor of naturally derived cosmetics. ● Tragtasche für eine Firma, die Naturkosmetik herstellt und vertreibt. ▲ Sac créé pour un fabricant et distributeur de produits de beauté naturels.

PAGE 189, image 427 ART DIRECTOR/DESIGNER: David Powell AGENCY: Gensler & Associates CLIENT: Stephanie Leigh's COUNTRY: USA PRINTER: Bagcraft ■ This shopping bag was created for an office tower specialty food retailer; the design objective was to be functional and promotional. ● Tragtasche für ein Delikatessengeschäft in einem Bürohaus. ▲ Sac créé pour une épicerie fine sise dans un bâtiment administratif.

PAGE 189, image 428 DESIGNERS: Alan Chan, Peter Lo ARTIST/ILLUSTRATOR: Peter Lo AGENCY: Alan Chan Design Company CLIENT: Hong Kong Tourist Association COUNTRY: Hong Kong ■ Inspired by the neon signs now typical of Hong Kong, the overlapping images of both Roman letters and Chinese calligraphy were used to re-create and project the city's distinctive qualities. ● Inspiriert von den für Hongkong typischen Neon-Reklamen, dienten die überlappenden lateinischen Buchstaben und chinesischen Schriftzeichen dazu, die besonderen Charakter der Stadt zum Ausdruck zu bringen. ▲ Inspirées des enseignes lumineuses typiques de Hong Kong, les calligraphies latine et chinoise qui se superposent ont été utilisées pour conférer un caractère unique à la métropole.

PAGE 189, image 429 ART DIRECTOR: Brian Deputy DESIGNER: Dave Gath ARTIST/ILLUSTRATOR: Joe Spencer AGENCY: Orbit City Art Company CLIENT: Hanna-Barbera COUNTRY: USA PRINTER: Spectrum Printing ■ This promotional shopping bag was created for the client's new animated series, "The Real Adventures of Johnny Quest." ● Tragtasche für eine Animations-Serie des Kunden mit dem Titel «Die wahren Abenteur von Johnny Quest». ▲ Sac créé pour promouvoir une série animée intitulée «Les Vraies Aventures de Johnny Quest».

PAGE 189, image 430 ART DIRECTOR: Jack Anderson DESIGNERS: Jack Anderson, Cliff Chung PRODUCT PHOTOGRAPHER: Tom McMackin ARTIST/ILLUSTRATOR: David Bates AGENCY: Hornall Anderson Design Works, Inc. CLIENT: University Villages PRINTER: Unisource TYPEFACE: Custom PAPER: Kraft stock ■ This shopping bag was created for a shopping complex named University Village. ● Tragtasche für einen Ladenkomplex mit dem Namen University Village. ▲ Sac pour un complexe commercial du nom d'University Village.

PAGE 189, image 431 ART DIRECTOR: Supon Phornirunlit DESIGNER: Steve Morris PRODUCT PHOTOGRAPHER: Oi Jakrarat Veerasarn AGENCY: Supon Design Group CLIENT: Museum of Junk COUNTRY: Singapore PRINTER: BKK Press TYPEFACE: New Baskerville PAPER: Kraft ■ Department store shopping bag. ● Tragtasche für ein Warenhaus. ▲ Sac conçu pour un grand magasin.

PAGE 189, image 432 ART DIRECTOR: Antonio Romano PRODUCT PHOTOGRAPHER: Giuseppe Maria Fadda AGENCY: Area Strategic Design CLIENT: (INCA) Instituto Nazionale Confederale di Assistenza COUNTRY: Italy TYPEFACE: Bodoni PAPER: Kraft ■ The packaging is for a conference held by a union-related company which defends workers rights. ● Tragtasche für eine Tagung des Instituto Nazionale Confederale di Assistenza, eine der Gewerkschaft angeschlossene Organisation, die sich für die Rechte der Arbeiter einsetzt. ▲ Emballage pour une conférence de l'INCA une organisation syndicale qui défend les droits des travailleurs.

PAGE 189, image 433 ART DIRECTOR: Antonio Romano PRODUCT PHOTOGRAPHER: Giuseppe Maria Fadda AGENCY: Area Strategic Design CLIENT: Farmindustria COUNTRY: Italy TYPEFACE: Bodoni PAPER: Kraft ■ This packaging was created for a *pharmaceutical company.* ■ *Tragtasche für ein pharmazeutisches Unternehmen.* ▲ *Sac créé pour une société pharmaceutique.*

PAGE 190, *image 434-435* ART DIRECTOR: *Antonio Romano* PRODUCT PHOTOGRAPHER: *Giuseppe Maria Fadda* AGENCY: *Area Strategic Design* CLIENT: *Berni* COUNTRY: *Italy* PRINTER: *G&G* TYPEFACE: *Bodoni Condensed* PAPER: *Kraft* ■ *This packaging was created for an upscale street-fashion retailer.* ● *Auftraggeber dieser Tragtasche ist ein Geschäft, das Strassenkleidung der gehobenen Preisklasse führt.* ▲ *Emballage créé pour une boutique de mode haut de gamme spécialisée dans les tenues de ville.*

PAGE 191, *image 436* CREATIVE DIRECTOR: *Peter Levine* DESIGNER: *Kim Tyska* PRODUCT PHOTOGRAPHER: *Andrew Bordwin* AGENCY: *Desgrippes Gobé & Associates* CLIENT: *Ann Taylor Loft* PAPER: *Recycled Kraft* ■ *The client's identity is intended to appeal to young women just beginning their careers. The shopping bag is designed to be simple, unpretentious and natural. However, it retains Ann Taylor's ink blue color and therefore fits into the heritage of the brand.* ● *Der Auftritt des Kunden soll junge Frauen ansprechen, die am Anfang ihrer beruflichen Karriere stehen. Die Tragtasche ist deshalb einfach, schlicht und natürlich. Durch Einbeziehung des für die Marke Ann Taylor typischen Tintenblaus fügt sie sich in das Gesamterscheinungsbild ein.* ▲ *L'identité du client s'adresse à des jeunes femmes qui débutent une carrière professionnelle. Le sac au design simple, naturel et discret reprend la couleur bleue typique de la marque Ann Taylor.*

PAGE 192, *image 437* ART DIRECTORS: *Paul Caldera, Doreen Caldera* DESIGNERS: *Tim Fisher, Bart Welch* AGENCY: *Caldera Design* CLIENT: *Deckers Outdoor Corporation* COUNTRY: *USA* ■ *This sandal was invented and is manufactured by Grand Canyon Colorado River guide, Mark Thatcher. The packaging system was designed to reinforce the authenticity of the company's heritage.* ● *Die Sandale wurde von Mark Thatcher, Grand-Canyon-Colorado-River-Führer erfunden und hergestellt. Das Verpackungssystem soll die Authentizität der Erzeugnisse der Herstellerfirma unterstreichen.* ▲ *Cette sandale a été conçue et fabriquée par Mark Thatcher, guide au Grand Canyon Colorado River. L'emballage doit souligner l'authenticité des produits de la société.*

PAGE 192, *image 438* ART DIRECTOR: *Steve Coleman* DESIGNER: *Chris De Lisen* AGENCY: *Elton Ward Design* CLIENT: *Dunlop Footwear* COUNTRY: *Australia* ■ *This brand identity design and shoe box design adaptation was developed for the K.T. Walker brand.* ● *Gestaltung der Marke und Verpackung für K.T. Walker-Schuhe.* ▲ *Identité de marque et emballage pour les chaussures K.T. Walker.*

PAGE 193, *images 439, 440* ART DIRECTOR: *Keith Steimel* DESIGNERS: *Paul McDowall, Jessica Valencia* ARTIST/ILLUSTRATOR: *Paul McDowall* AGENCY: *Cornerstone* CLIENT: *Converse* COUNTRY: *USA* ■ *The objective was to give life to the color-changing liquid-crystal technology used in the sneakers. The containment box acts as a vault for the "blob" which invades the sneakers.* ● *Hier ging es um die durch Flüssigkristall bewirkte Farbveränderung der Schuhe. Die Schachtel ist quasi ein Verliess, damit der Geist, der sich in die Schuhe eingenistet hat, nicht entweichen kann.* ▲ *Il s'agissait de mettre en valeur le changement de couleur des chaussures de sport obtenu grâce à des cristaux liquides.*

PAGE 194, *image 441* ART DIRECTOR: *José Serrano* DESIGNERS: *Mike Brower, José Serrano* PHOTOGRAPHER: *Carl Van der Shuit* ILLUSTRATOR: *Tracy Sabin* AGENCY: *Mires Design, Inc.* CLIENT: *Voit Sports, Inc.* COUNTRY: *USA* ■ *This packaging was designed to appeal to young consumers. Through the graphics and photography, the packaging is also intended to convey the product's versatility for both indoor and outdoor use.* ■ *Graphik und Photographie der Verpackung, die sich an eine junge Kundschaft richtet, sollen zum Ausdruck bringen, dass sich das Produkt für den Gebrauch im Hause sowie im Freien eignet.* ▲ *Le graphisme et la photographie de cet emballage qui s'adresse à de jeunes consommateurs doivent véhiculer l'idée que le produit peut être utilisé aussi bien à l'intérieur qu'à l'extérieur.*

PAGE 195, *image 442* ART DIRECTOR: *Tim Larsen* DESIGNERS: *David Schultz, Sascha Boecker* PHOTOGRAPHER: *Curtis Johnson (Arndt Studios)* AGENCY: *Joe Paczkowski (Ripsaw Photography)* AGENCY: *Larsen Design + Interactive* CLIENT: *First Team Sports* COUNTRY: *USA* PRINTER: *Willamette Industries* TYPEFACE: *Orator, Rockwell* PAPER: *Pre-Paint Corrugate* ■ *This in-line skate packaging was created to introduce new and redesigned products.* ● *Hier ging es um die Einführung neuer bzw. verbesserter In-Line Skates.* ▲ *Emballage de patins à roulette «inline» créé pour le lancement de nouveaux produits et de produits relookés.*

PAGE 196, *image 443* ART DIRECTOR: *Mary Scott* DESIGNERS: *Camille Favilli, Carrie Dobbel* ARTISTS/ILLUSTRATORS: *Martin Ledyard, Bruce Morser* AGENCY: *Maddocks & Company* CLIENT: *Bell Helmets* COUNTRY: *USA* PRINTER: *Color Box* TYPEFACE: *Adobe Garamond, Gill Sans* PAPER: *Laminated E-Flute, Stock 70# Coated Label Stock* ■ *This packaging was created for the mass- and pro- sportswear markets.* ● *Diese Verpackung ist vor allem für Kaufhäuser und grosse Sportgeschäfte bestimmt.* ▲ *Cet emballage est destiné avant tout aux grandes surfaces et aux magasins de sport.*

PAGE 196, *images 444, 445* ART DIRECTOR: *Mary Lewis* DESIGNER: *Martin McLoughlin* PRODUCT PHOTOGRAPHER: *Robin Broadbent* AGENCY: *Lewis Moberly* CLIENT: *Halfords Ltd* COUNTRY: *United Kingdom* ■ *The packaging design and name*

were created to appeal to a young, style-conscious audience and to convey the product's use of state-of-the-art technology. ● *Verpackung und Name des Produktes sollen ein junges, stilbewusstes Publikum ansprechen und das hohe technologische Herstellungsniveau zum Ausdruck bringen.* ▲ *L'emballage et le nom du produit s'adressent à un public cible jeune et sont sensés mettre en valeur la technologie de pointe utilisée pour la fabrication de ce produit.*

PAGE 197, *images 446, 447* ART DIRECTOR: *Arnold Goodwin* DESIGNERS: *Arnold Goodwin, Lisa Kosinar, Jim Costello* PHOTOGRAPHER: *Jim White* ARTIST/ILLUSTRATOR: *Mark Summers* AGENCY: *Stanaszek Goodwin Design* CLIENT: *Beta Sports, Inc.* COUNTRY: *USA* PRINTER: *Ambassador Press* TYPEFACE: *Copperplate Gothic, Garamond Ital.* PAPER: *Corrugated & 10 pt. Coated* ■ *This direct market packaging was developed to promote product retention after delivery, support a high-middle price point, quickly give the customer a visual communication of product values and attention to detail, and create and support brand awareness.* ● *Bei dieser Verpackung für den Direktversand ging es um die Darstellung eines Produktes der gehobenen mittleren Preisklasse, seines Wertes und der sorgfältigen Herstellung sowie um eine eindrucksvolle Präsentation der Marke.* ▲ *Cet emballage de V.P.C. devait inciter le client à garder le produit après livraison, mettre en valeur le produit qui figure dans une catégorie de prix moyenne et faire ressortir ses qualités et sa fabrication minutieuse tout en contribuant à la notoriété de la marque.*

PAGE 198, *images 448, 449* ART DIRECTOR: *Peter Moore* DESIGNERS: *Rolf Heidemeier, Susanne Kubler* CLIENT: *Adidas* ■ *New design of the packaging with environment-friendly, recycled material.* ● *Neugestaltung der Verpackung mit umweltfeundlichem, wiederverwertetem Material.* ▲ *Nouvel emballage réalisé avec des matériaux recyclés et respectueux de l'environnement.*

PAGE 199, *image 450* ART DIRECTORS: *Paul Caldera, Doreen Caldera* DESIGNERS: *Tim Fisher, Bart Welch* PHOTOGRAPHER: *Bob Carey* AGENCY: *Caldera Design* CLIENT: *Deckers Outdoor Corporation* COUNTRY: *USA* ■ *The packaging system was designed to build brand identity and communicate product feature benefits to both domestic and international consumers of bicycle accessories.* ● *Das Verpackungssystem sollte den Markenauftritt unterstützen und den Käufern von Fahrradzubehör im In- und Ausland die vorteilhaften Produkteigenschaften vermitteln.* ▲ *Système d'emballage conçu pour promouvoir l'identité de la marque et présenter les qualités et avantages du produit aux acheteurs d'accessoires pour cycles.*

PAGE 199, *image 451* ART DIRECTOR: *José Serrano* DESIGNERS: *José Serrano, Mike Brower* ARTIST/ILLUSTRATOR: *Tracy Sabin* AGENCY: *Mires Design Inc.* CLIENT: *Voit Sports, Inc.* COUNTRY: *USA* ■ *Voit Sports and Body Glove created a surf fin designed specifically for body surfing. The packaging was designed to have a fresh, young look which is attractive to the ocean-going market.* ● *Voit Sports und Body Glove haben eine Surf-Flosse entwickelt, die sich besonders zum Body Surfen eignet. Die Verpackung sollte jung und frisch wirken und damit die spezielle Zielgruppe ansprechen.* ▲ *Voits Sports et Body Glove ont créé une nageoire de surf conçue tout spécialement pour le «body surfing». L'emballage devait avoir un look jeune et frais à l'image du public cible.*

PAGE 199, *image 452* ART DIRECTOR: *José Serrano* DESIGNERS: *José Serrano, Mike Brower* ARTIST/ILLUSTRATOR: *Tracy Sabin* AGENCY: *Mires Design Inc.* CLIENT: *Ektelon* COUNTRY: *USA* ■ *Ektelon makes shoes, apparel, gear bags and accessories for the racquetball enthusiast. The re-design of their gear bag and racquet graphics gave the merchandise a fresh, new look.* ● *Ektelon stellt Schuhe, Kleidung und Zubehör sowie Zubehörtaschen für Racquetball-Liebhaber her. Die Überarbeitung ihrer Tasche und der Raquet-Graphik gab den Artikeln ein frisches, neues Aussehen.* ▲ *Ekelton fabrique des chaussures, des vêtements, des sacs de sports et des accessoires pour les passionnés de jeux de raquette. La nouvelle conception du sac de sport et du graphisme de la raquette donne une impression de fraîcheur.*

PAGE 199, *image 453* ART DIRECTORS: *Dana Lytle, Kevin Wade* DESIGNERS: *Dana Lytle, Raelene Mercer* PRODUCT PHOTOGRAPHER: *Mark Salisbury* PRODUCT PHOTOGRAPHER: *Sutter Photography* ARTIST/ILLUSTRATOR: *Dana Lytle* AGENCY: *Planet Design Company* CLIENT: *Graber USA* COUNTRY: *USA* PRINTER: *Philipp Lithography, Printon Engravers, Castle Rock Containers* TYPEFACE: *OCRA* PAPER: *70 lb. Champion* ■ *Outback is a line of bicycle racks sold through mass-merchandisers. The packaging was designed to convey a feeling of the established brand while at the same time communicating all the requisite mass-merchandiser information.* ● *Outback ist eine Linie von Gepäckträgern für Fahrräder, die als Massenware verkauft werden. Die Verpackung musste den Eindruck einer etablierten Marke vermitteln und gleichzeitig all die für die entsprechenden Vertriebskanäle erforderlichen Produktinformationen enthalten.* ▲ *Outback est une gamme de porte-bagages pour cycles destinée à la grande distribution. L'emballage devait donner l'impression que la marque est bien établie et communiquer en même temps toutes les informations sur le produit nécessaires aux différents distributeurs.*

PAGE 199, *image 454* CREATIVE DIRECTOR: *Michael Jager* ART DIRECTOR: *David Covell* DESIGNER: *Dan Sharp* AGENCY: *Jager Di Paola Kemp Design* CLIENT: *Burton Snowboards* COUNTRY: *USA* PRINTER: *General Press* TYPEFACE: *Rotis Sans, AG Book Rounded (modified)* ■ *The boot boxes were designed to present the attitude of each boot and the style of riding for which each boot was designed, as well as act as its own point of sale item. All technical information and riding specifics are* *available to compare and contrast with other boots.* ● *Diese Stiefelschachteln geben Auskunft über die Eigenschaften der Stiefel und den Snowboard-Typ, für den sie entwickelt wurden, so dass sie sich leicht mit anderen Stiefeln vergleichen lassen. Gleichzeitig dienen sie als Ladendisplays.* ▲ *Ces boîtes prévues pour contenir des bottes de snowboard donnent des informations sur les caractéristiques du produit et leur utilisation finale afin de pouvoir les comparer à d'autres bottes. Elles servent également de présentoir de magasin.*

PAGE 199, *image 455* ART DIRECTOR: *Neil Powell* DESIGNER/ARTIST/ILLUSTRATOR: *Missy Wilson* AGENCY: *Duffy Design* CLIENT: *Duke Design Inc.* COUNTRY: *USA* ■ *This design was created to introduce the product into a larger market.* ● *Aufgabe dieser Verpackung ist es, das Produkt in einem grösseren Markt einzuführen.* ▲ *Ce design a été conçu pour élargir les canaux de distribution de ce produit.*

PAGE 199, *images 456–458* ART DIRECTOR: *Peter Moore* DESIGNERS: *Rolf Heidemeier, Susanne Kubler* CLIENT: *Adidas* ■ *New design of the package with environment-friendly recycled material.* ● *Neugestaltung der Verpackung mit umweltfreundlichen Recycling-Material.* ▲ *Nouveau design de l'emballage réalisé avec des matériaux recyclés respectueux de l'environnement.*

PAGE 200, *image 459* ART DIRECTORS: *Fritz Haase, Harald Schweers* DESIGNERS: *Andreas Wilhelm, Claudia Buschmann-Tunsch* AGENCY: *Atelier Haase & Knels* CLIENT: *Stanwell Vertriebs GmbH* COUNTRY: *Germany* PRINTER: *Bohlmeier & Co. Kirchlengern, Germany* ■ *Package design for the launch of a range of cigarillos.* ● *Verpackungsgestaltung für die Neueinführung von Zigarillos.* ▲ *Design d'emballage pour le lancement de cigarillos.*

PAGE 201, *image 460* ART DIRECTORS: *Holger Sinn, Harald Schweers* DESIGNERS: *Holger Sinn, Ulf Nawrot* ILLUSTRATOR: *Ulf Nawrot* AGENCY: *Atelier Haase & Knels* CLIENT: *Stanwell Vertriebs GmbH* COUNTRY: *Germany* PRINTER: *Bohlmeier & Co. Kirchlengern, Germany* ■ *New packaging design of a range for cigarillos and cigars. Neuentwicklung einer Packungsserie für Zigarillos und Zigarren.* ▲ *Nouveau design d'une gamme d'emballages pour cigarillos et cigares.*

PAGE 202, *image 461* ART DIRECTOR: *Keith Steimel* DESIGNER: *Paul McDowall* ARTIST/ILLUSTRATOR: *Tom Montini* AGENCY: *Cornerstone* CLIENT: *Moonlight Tobacco Co.* COUNTRY: *USA* ■ *This packaging was designed to create an identity which mimics the wide gauge features of the cigarette brand* ● *Diese Verpackung sollte eine Produktidentität schaffen, die die typischen Eigenschaften der Zigarettenmarke reflektiert.* ▲ *Emballage conçu dans le but de créer une identité qui reprend les caractéristiques de la marque de cigarettes.*

PAGE 203, *image 462* ART DIRECTOR: *Keith Steimel* DESIGNER: *Paul McDowall* AGENCY: *Cornerstone* CLIENT: *Moonlight Tobacco Co.* COUNTRY: *USA* ■ *The task was to create a very upscale, sexy, comforting look for a honey-toasted cigarette.* ● *Der Auftritt dieser Zigarette mit Honigaroma sollte sehr anspruchsvoll, sexy und genussvoll wirken.* ▲ *L'image de cette cigarette au goût de miel devait être à la fois sexy et haut de gamme, et donner l'impression d'un sentiment de bien-être.*

PAGE 203, *image 463* ART DIRECTOR: *Keith Steimel* DESIGNERS: *Paul McDowall, Keith Steimel* ARTIST/ILLUSTRATOR: *Tom Montini* AGENCY: *Cornerstone* CLIENT: *Moonlight Tobacco Co.* COUNTRY: *USA* ■ *Usually one would expect a very sexy design for a honey-toasted product. The design strategy was to contradict that trend and go for a very "in your face" look. In an attempt to defy industry norms, the package has no text logo.* ● *Allgemein erwartet man von einem Produkt mit Honigaroma einen sexy Auftritt und von einer Zigarettenpackung einen*

Schriftzug. Hier entschied man sich, den allgemeinen Erwartungen bzw. dem Trend nicht zu entsprechen. ▲ *D'ordinaire, on s'attend à un design sexy pour un produit au goût de miel et on a l'habitude de voir figurer des caractères ou un logo sur un paquet de cigarettes. La stratégie appliquée au design de ce paquet défie les normes industrielles et va à l'encontre de toutes les tendances.*

PAGE 204, 205, *images 464, 466* ART DIRECTOR: *Red Maxwell* DESIGNER: *Deborah Pingitore* PRODUCT PHOTOGRAPHER: *Steve Cash* AGENCY: *International Imaging Inc.* CLIENT: *R.J. Reynolds International B.V.* COUNTRY: *Germany* TYPEFACE: *In-house* ■ *This redesigned packaging was created for Yves Saint Laurent premium cigarettes.* ● *Neugestaltung der Verpackung für Zigaretten von Yves Saint Laurent.* ▲ *Nouvel emballage pour des cigarettes Yves Saint Laurent de qualité supérieure.*

PAGE 204, *image 465* ART DIRECTOR: *Keith Steimel* DESIGNER: *Paul McDowall* AGENCY: *Cornerstone* CLIENT: *Moonlight Tobacco Co* COUNTRY: *USA* ■ *This packaging was designed to create an identity which depicts the qualities of this east-west blend of tobaccos.* ● *Bei dieser Zigarettenpackung ging es um die Darstellung der speziellen Tabakmischung.* ▲ *Pour ce paquet de cigarettes, il s'agissait de créer une identité soulignant les qualités de ce mélange de tabacs.*

PAGE 206, *image 467* ART DIRECTOR/DESIGNER: *Andreas Wilhelm* AGENCY: *Atelier Haase & Knels* CLIENT: *Stanwell Vertriebs GmbH* COUNTRY: *Germany* PRINTER: *Herbert Klann Metall, Blechfabrik GmbH* ■ *Package design for the launch of a range of pipe tobacco.* ● *Verpackungsgestaltung für die Lancierung von Pfeifentabak bester Qualität.* ▲ *Emballage créé pour le lancement d'un tabac pour pipe de qualité supérieure.*

PAGE 207, *images 468, 469* ART DIRECTORS: *Sibylle Haase, Harald Schweers* DESIGNERS: *Harald Schweers, Andreas Wilhelm* AGENCY: *Atelier Haase & Knels* CLIENT: *Stanwell Vertriebs GmbH* COUNTRY: *Germany* PRINTER: *Herbert Klann Metall, Blechfabrik GmbH* ■ *Package design for the launch of a range of pipe tobacco.* ● *Verpackungsgestaltung für die Lancierung von Pfeifentabak bester Qualität.* ▲ *Emballage créé pour le lancement d'un tabac pour pipe de qualité supérieure.*

PAGE 207, *image 470* ART DIRECTORS: *Sibylle Haase, Harald Schweers* DESIGNERS: *Thomas Meyer, Regina Spiekermann* ILLUSTRATOR: *Dirk Bergner* AGENCY: *Atelier Haase & Knels* CLIENT: *Stanwell Vertriebs GmbH* COUNTRY: *Germany* PRINTER: *Herbert Klann Metall, Blechfabrik GmbH* ■ *Package design for the launch of a range of pipe tobacco.* ● *Verpackungsgestaltung für die Lancierung von Pfeifentabak bester Qualität.* ▲ *Emballage créé pour le lancement d'un tabac pour pipe de qualité supérieure.*

PAGE 207, *image 471* ART DIRECTORS: *Sibylle Haase, Harald Schweers* DESIGNERS: *Katia Hirschfelder, Regina Spiekermann* AGENCY: *Atelier Haase & Knels* CLIENT: *Stanwell Vertriebs GmbH* COUNTRY: *Germany* PRINTER: *Herbert Klann Metall, Blechfabrik GmbH* ■ *Package design for the launch of a range of pipe tobacco.* ● *Verpackungsgestaltung für die Lancierung von Pfeifentabak bester Qualität.* ▲ *Emballage créé pour le lancement d'un tabac pour pipe de qualité supérieure.*

PAGE 208, *image 472* ART DIRECTORS: *Sibylle Haase, Harald Schweers* DESIGNERS: *Martina Lüllich, Regina Spiekermann* AGENCY: *Atelier Haase & Knels* CLIENT: *Stanwell Vertriebs GmbH* COUNTRY: *Germany* PRINTER: *Herbert Klann Metall, Blechfabrik GmbH* ■ *Package design for the launch of a range of pipe tobacco.* ● *Verpackungsgestaltung für die Lancierung von Pfeifentabak bester Qualität.* ▲ *Emballage créé pour le lancement d'un tabac pour pipe de qualité supérieure.*

C R E A T I V E   D I R E C T O R S • A R T   D I R E C T O R S • D E S I G N E R S

CLIENTS

GRAPHIS BOOKS

BOOK ORDER FORM: USA, CANADA, SOUTH AMERICA, ASIA, PACIFIC

BOOKS	ALL REGIONS
☐ BLACK & WHITE BLUES (HARDCOVER)	US$ 69.95
☐ BLACK & WHITE BLUES (PAPERBACK)	US$ 45.95
☐ GRAPHIS ADVERTISING 97	US$ 69.95
☐ GRAPHIS ALTERNATIVE PHOTOGRAPHY 95	US$ 69.95
☐ GRAPHIS ANNUAL REPORTS 5	US$ 69.95
☐ GRAPHIS BOOK DESIGN	US$ 75.95
☐ GRAPHIS BROCHURES 2	US$ 75.00
☐ GRAPHIS CORPORATE IDENTITY 2	US$ 75.95
☐ GRAPHIS DESIGN 97	US$ 69.95
☐ GRAPHIS EPHEMERA	US$ 75.95
☐ GRAPHIS FINE ART PHOTOGRAPHY	US$ 85.00
☐ GRAPHIS INFORMATION ARCHITECTS	US$ 49.95
☐ GRAPHIS MUSIC CDS	US$ 75.95
☐ GRAPHIS NUDES (PAPERBACK)	US$ 39.95
☐ GRAPHIS PACKAGING 7	US$ 75.00
☐ GRAPHIS PHOTO 96	US$ 69.95
☐ GRAPHIS POSTER 96	US$ 69.95
☐ GRAPHIS PRODUCTS BY DESIGN	US$ 69.95
☐ GRAPHIS SHOPPING BAGS	US$ 69.95
☐ GRAPHIS TYPOGRAPHY 1	US$ 69.95
☐ GRAPHIS TYPE SPECIMENS	US$ 49.95
☐ **GRAPHIS PAPER SPECIFIER SYSTEM (GPS)** ** ADD $30 SHIPPING/HANDLING FOR GPS	US$ 495.00
☐ HUMAN CONDITION	US$ 49.95
☐ SHORELINE	US$ 85.95
☐ WATERDANCE (PAPERBACK)	US$ 24.95
☐ WORLD TRADE MARKS 100 YRS.(2 VOL. SET)	US$ 250.00

NOTE! NY RESIDENTS ADD 8.25% SALES TAX

☐ CHECK ENCLOSED (PAYABLE TO GRAPHIS)
(US$ ONLY, DRAWN ON A BANK IN THE USA)

USE CREDIT CARDS (DEBITED IN US DOLLARS)

☐ AMERICAN EXPRESS ☐ MASTERCARD ☐ VISA

CARD NO. EXP. DATE

CARDHOLDER NAME

SIGNATURE

(PLEASE PRINT)

NAME

TITLE

COMPANY

ADDRESS

CITY

STATE/PROVINCE ZIP CODE

COUNTRY

SEND ORDER FORM AND MAKE CHECK PAYABLE TO:
GRAPHIS INC.,
141 LEXINGTON AVENUE, NEW YORK, NY 10016-8193, USA

GRAPHIS MAGAZINE

SUBSCRIBE TO GRAPHIS: USA, CANADA, SOUTH AMERICA, ASIA, PACIFIC

MAGAZINE	USA	CANADA	SOUTHAMERICA/ ASIA/PACIFIC
☐ ONE YEAR (6 ISSUES)	US$ 89.00	US$ 99.00	US$ 125.00
☐ TWO YEARS (12 ISSUES)	US$ 159.00	US$ 179.00	US$ 235.00
☐ AIRMAIL SURCHARGE (6 ISSUES)	US$ 59.00	US$ 59.00	US$ 59.00

☐ ONE YEAR (6 ISSUES) US$ 59.00
FOR STUDENTS WITH COPY OF VALID STUDENT ID AND PAYMENT WITH ORDER

☐ CHECK ENCLOSED ☐ PLEASE BILL ME

USE CREDIT CARDS (DEBITED IN US DOLLARS)

☐ AMERICAN EXPRESS

☐ MASTERCARD

☐ VISA

CARD NO. EXP. DATE

CARDHOLDER NAME

SIGNATURE

(PLEASE PRINT)

NAME

TITLE

COMPANY

ADDRESS

CITY

STATE/PROVINCE ZIP CODE

COUNTRY

SERVICE BEGINS WITH ISSUE THAT IS CURRENT WHEN ORDER IS PROCESSED.

SEND ORDER FORM AND MAKE CHECK PAYABLE TO:
GRAPHIS INC.,
141 LEXINGTON AVENUE, NEW YORK, NY 10016-8193, USA

(C9B0A)

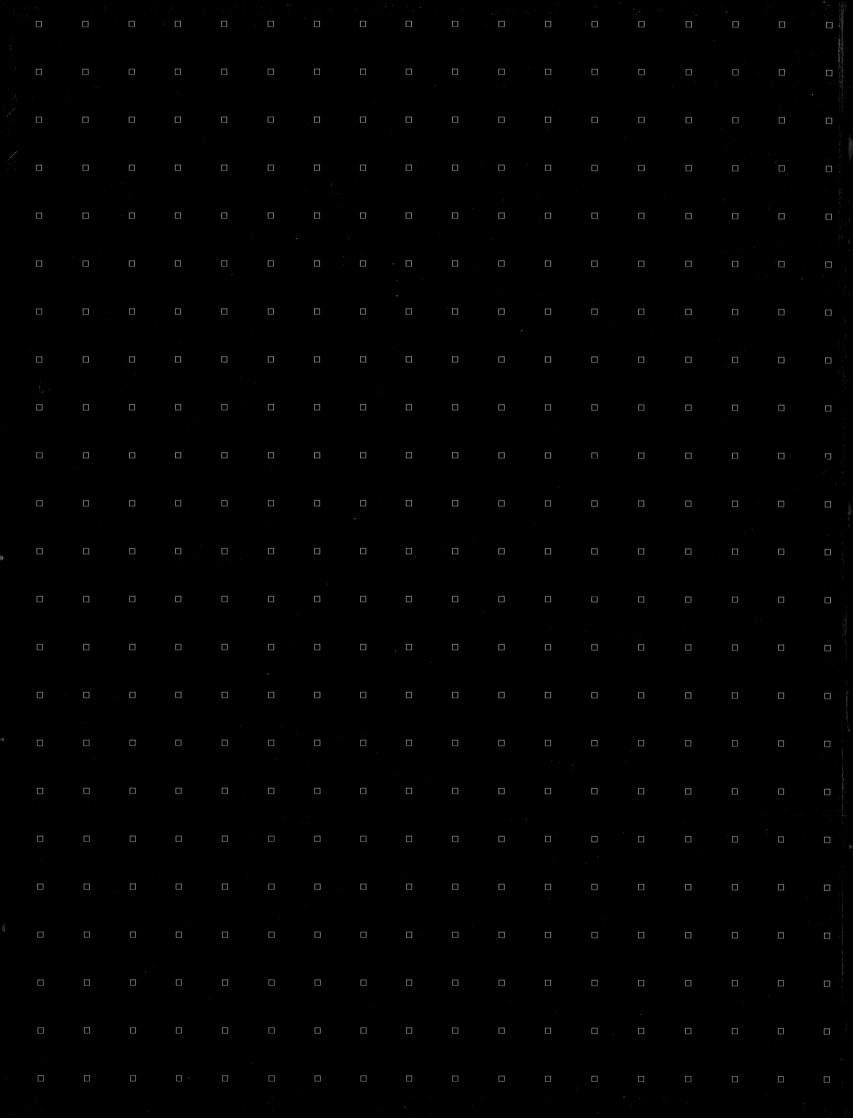